No Prize
for Pessimism

No Prize for Pessimism

/ Letters from a messy
    tech optimist

< **SAM SCHILLACE**

Microsoft's Deputy Chief
Technology Officer
Inventor of Google Docs

an imprint of Microsoft

© 2024 Microsoft Corporation. All rights reserved.

Published in the United States by 8080 Books,
an imprint of Microsoft Corporation
1 Microsoft Way, Redmond, WA 98052

https://aka.ms/8080books

Cover Design by Shyla Lindsey
Typesetting by Mariner Media
In association with Partners Media

To have a conversation with this book,
visit www.schillace.com/letters

ISBN 9798991762304 (Print – 8080 Books)
ISBN 9798991762311 (eBook – 8080 Books)

MICROSOFT, 8080 BOOKS, and the 8080 BOOKS Logo
are trademarks of Microsoft Corporation

For my wife, Angela, who is always willing to have ice cream in the rain, ride the rollercoaster, and have fun adventures with me.

Table of Contents

Preface ix

Chapter 1 1
I Choose Optimism

Chapter 2 23
Google Docs: A Story of Invention

LETTERS FROM A MESSY TECH OPTIMIST

Chapter 3 45
Mindset: Approaching Disruption and Innovation

Chapter 4 81
Technical Philosophy

Chapter 5 97
AI and Products

Chapter 6 151
Product Design

Chapter 7 — 169
Humility in Leadership … and the Prima Donna Death Spiral

Chapter 8 — 181
Systems, Organizations, and Incentives

Chapter 9 — 203
Career Advice

Chapter 10 — 235
Teatime with an Alien

Chapter 11 — 247
The Schillace Laws

Chapter 12 — 259
A Watch List for Future Innovators

Chapter 13 — 281
It's Your Turn to Embrace Optimism in the Age of AI

Acknowledgments — 297

Preface

Whether you are an entrepreneur, programmer, engineer, or any other professional, you get older, and the next generation ages into the roles you once occupied. For generations, the young have dreamed bigger and better dreams than those before them, creating a virtuous circle of optimistic beginners and seasoned mentors. One of my favorite things is speaking to new engineers and students. They always have fresh perspectives and a disruptive energy that challenges the way things are.

But in recent years, I've noticed a disturbing trend. The young people entering the workforce increasingly eschew optimism and embrace pessimism. Blogger and activist Matt Ball writes that he used to believe optimists were naïve and pessimists were smart. "Pessimism seemed like an essential feature of a scientist." Now he channels a colleague, Max Roser, who states: "The world is awful. The world is much better. The world can be much better." In other words, it's possible to hold several truths at once, and recognizing that things have gotten better enables us to see that a better world is possible. Pessimists, Ball notes, are not interested in solutions; they've already given up. Yet pessimists seem to be everywhere. According to the Pew Research Center, about 7 in 10 Americans think young adults today have a harder time than their parents' generation when it comes to saving for the future (72%), paying for college (71%), and buying a home (70%). In an op-ed in 2024, Fareed Zakaria questioned why so many Americans have "a profound sense of despair."

There have always been pessimists and naysayers, of course. But they are usually found in the older generation, with more at stake in the status quo. This engine of disruption and innovation is essential. Everything we have, from the invention of fire to AI today, was initially "against the grain." Heresy is the lifeblood of civilization, and the new generation is usually its standard bearer. Even more importantly, in our current moment, the global demographic challenges mean there will be more pressure on this new generation to innovate and create. The pessimistic mindset is dangerous!

So what's with this inversion of sentiment from optimist to pessimist? Is it possible that the next generation will be less innovative than the last? If the rise of American tech has taught us anything, it is that there is no prize for being pessimistic and right, but there is one for being optimistic and correct (and usually contrarian!). Mindset is critical. New ideas are tough and, as we will see, nothing is evident at first. Entrepreneurship is about moving forward when everyone thinks you're wrong, being optimistic, and holding on to that conviction until your "radical idea" becomes the obvious thing everyone has always agreed on. The core of this mindset is a willingness to experiment, explore, and fix problems. Mess and uncertainty are at the heart of it. Solutions aren't apparent until you start working on the problem, and sometimes not even then! Sometimes, they come from unexpected directions while you are exploring something else.

In 2012, while at the cloud computing company Box, I began writing brief letters to the young engineers on our staff. I'd sit down every Sunday afternoon and reflect on the past week and the week ahead. More than a decade later, I still write these Sunday Letters in part to prepare myself for the week ahead and in part to share lessons and observations (one nice side effect of this is that the regular commitment helps me be "present" during the week, looking for commonalities and systemic observations I can generalize and share). As the audience for these letters has grown through Substack and LinkedIn, I've been told they have another impact on readers, one I had not anticipated. They help bring readers into the mind of a programmer, showing how to decompose problems into first principles and patterns and move from there to small steps and tools to progress through the problem.

This book is my attempt to build on the thinking behind these *Sunday Letters*. I hope to do two things. First, I want to share the first principles I've learned and used to disrupt and innovate at Google, Microsoft, and Box. I want you to take some fundamental tools away from reading this.

Second, we will explore some of these ideas in the current moment of AI. Innovation and disruption are slippery things. Just like it's straightforward to see the path followed through the thicket when you turn around, it's much easier to see how something became important after the fact. But in the moment, all you see is the thicket. History is exciting, but only so much as it helps us in the current moment with the next set of uncertain tools and new problems.

Throughout my career as a software programmer, I've been as curious about solving challenging tech problems as I have about the conditions that produce innovation. I was fortunate enough to be part of the Google Docs story at the start, and we'll talk more about that shortly, but one of the striking things to me is how skeptical (and often actively opposing) folks were initially. I've also always been a tinkerer and mess-maker, and I've thought a lot about how process and mindset impact the act of creation.

How can we move from heuristics (or trial-and-error problem solving) to precision? What's the right level of mess and experiment versus focusing on a fixed path? One mistake we all make is to look at the current moment in isolation, as a single dot on a graph. In reality, everything is moving all the time. It's well known that humans struggle to model anything non-linear. Since most tech is exponential, we often don't have a good intuition for how it will evolve. But we must try to understand not just the single point, but the curve — not where we are but where we are going. Don't look at a single dot; look for the trend with many.

Everyone wants to be innovative and create the next big, disruptive thing. But how do we train ourselves to produce these ideas? Several writers have asked similar questions. Author Leander Kahney writes about Jony Ive, the design genius behind Apple's great products. In *The Creative Act: A Way of Being*, Rick Rubin, the co-founder of Def Jam Recordings, writes that everyone is a creator. His meditations on tuning in, source, practice, intention, great expectations, and how to choose offer proven inspiration to the artist in all of us. Similarly, my ruminations on embracing the slow and messy, the role of simplicity, having a beginner's mind, and how to receive new ideas are for the innovator in all of us.

There's a common misperception that the big ideas are self-evident when you see them. This can be true, but the big ideas are more often messy, impractical, and hard to distinguish from bad ideas. In some sense, they have to be — if they were apparent, either you're incredibly lucky (possible, but by definition unlikely) or everyone else would have seen them already. There's also a pattern we'll discuss later: The big ideas are often divisive: You get some very strong in favor and some very strong in opposition. We will show why this happens and how you can use it to spot these big ideas yourself.

Looking for a good, finished idea is pointless. It would be like mining and expecting to find finished jewelry. That's not how it works. Good miners can see potential in the raw materials. If you've ever seen a gemstone as it comes out of the ground, it usually looks like a pebble. Great ideas are like that, too.

You have to have similar skills to mine for good ideas. You have to be able to look past the complex challenges and unfinished bits and see the value inside them. The ability to look at something and insert *[what if?]* is a critical skill we will explore and develop. This central theme can help destroy a pessimistic mindset.

I've argued that optimism is essential for innovation, creativity, and good ideas. So, how might AI improve or impede this? You might be skeptical about AI or even have a list of well-reasoned concerns and challenges. It might be that AI won't work for everything people are trying now (almost certainly true). But, in my view, we will learn that the AI moment is very much like so many other disruptive moments in tech, and the mindset you bring to it is critical. Waiting for the answer to be obvious is like sitting on the sidelines of the game. The time to innovate with anything (including AI) is when it's uncertain.

The core idea of this book is that so many breakthroughs we take for granted today emerged from new ideas that initially seemed wrong. These ideas were laughed at. A research paper that lives in the cloud so multiple authors can collaborate simultaneously? A phone and camera in your pocket? A great car that runs on electricity? A flying taxi?

Humans, we, the people, are the starting point. We bring the optimism to innovate. What we'll see with AI is that it fits all of the patterns that will be described in this book — it's disruptive and uncomfortable, and people are forming narratives to challenge it, but it's such a powerful tool that we must see it objectively, with clear eyes. Let's insert *[what if?]*.

Don't be fooled by early mistakes, reporters, or others who make a living writing scare stories, or all the other flavors of "Why bother?" out there. AI is a fundamental shift that we will be working through for quite some time and, eventually, it will reorder virtually everything about the technical industry at some level.

The thorniest challenge in developing AI that is autonomous and truly collaborative is to express an intention such that a program can creatively find the solution without having to be told. The challenge also exists when developing young talent who can find a unique solution without specific directions. Sadly, the generation currently developing this tech too often needs this skill. The desire to succeed right away breeds a reluctance to experiment. We should teach computers to think like the most collaborative humans. In that case, we're reminded of what skills we need to cultivate in ourselves and the next generation of innovators.

The late Paul Allen, a co-founder of Microsoft, expressed it this way: "I choose optimism. I hope to be a catalyst not only by providing financial resources but also by fostering a sense of possibility: encouraging top experts to collaborate across disciplines, challenge conventional thinking, and figure out ways to overcome some of the world's hardest problems."

I, too, choose optimism.

Chapter 1

I Choose Optimism

I started writing software professionally way back in the "dark ages" of 1989. The Internet had quietly reached 100,000 host machines, and Intel had introduced its 80486. Even before that, I was starting to write code as a hobby, though I didn't really think of it as anything serious — just another in a long list of hobbies and experiments, along with trying to make dandelion wine, woodcarving, catching turtles, and generally exploring the world.

When I was in eighth grade, I saw my first personal computer, a PET, at a summer camp for so-called smart kids. Later on, in high school, one of my friends had an Apple II, and we played video games on it. I thought computers were fun and interesting, and I liked to mess around with them.

After some begging, my dad bought me a TRS-80 when I was about 12 (lucky kid!). I still have no idea why (Thanks, Dad!). And I did the same thing every kid did in that era: I copied programs by hand from the back of the *Dr. Dobbs Journal*, and learned to code by observing and trying things. That instinct — to just try things, learn what you can and iterate from whatever works, solving problems one at a time — has stayed with me my whole career. It's what most of this book is about: how to solve problems when there isn't much (if any) of a roadmap.

(As an aside, one of the first programs I copied out was Eliza — a very simple chat program that would mostly just echo back what you had said and ask a few questions. It is nothing special by today's standards, but it was magical to 12-year-old me back in Michigan. It's hard to describe the feeling of having been in this industry long enough so that we now have real versions of that early "chatbot" that can reason, ask complex questions, deal with ambiguity, be friends and advisers, and so much more. This is what I love about the tech industry: You get to live through and help create incredible change).

Back then, of course, there was no internet, there was barely any documentation other than things like the *Dr. Dobbs Journal*, and it was hard to even store and reuse programs. We had to load them on cassette tapes, a process that was slow and hard to get right. That was okay, you couldn't write much code anyway! The pixels on that early computer weren't even square! I remember

2   Sam Schillace

spending a lot of time trying to draw a straight line with them. It was all a mess, but I had fun exploring anyway.

None of this seemed super serious, or even possible, to the generation that came before us. They were mostly electrical engineers and mathematicians, and for them computers were not far removed from raw electronics, limited in function and all about precision. There was a lot of sexism in those days, too. You weren't considered a "macho man" if you liked the "pretty pixels" of the new PCs. "Real" programmers sweated with mainframes or minis, debugged with oscilloscopes, and did serious work for serious people, slowly, usually in COBOL. There wasn't anything like the kind of tech industry or career path that we have now. Everything was slow, dull, opaque, and hard.

The old guys thought this PC thing was dumb. It didn't do any of the things like true multitasking that their "big iron" did. PCs were smaller — less memory, less capable CPUs, and they didn't have all the software on them that the old guard was used to. They didn't even have proper hard disks at first. They had these weird new operating systems like DOS, and programming languages like BASIC.

From their perspective, the PC was clearly inferior, and they let us know it. Their pessimism was meant to drown our optimism. But my friends and I could immediately see the power of this new model, and we had the imagination to see what might happen when everyone had one. I don't think any of us had any idea that we'd eventually have lots of computers all connected to each other on the internet, or that we'd carry capable computers around in our pockets all the time. That would have seemed SO COOL to us at the time, and some visionaries were seeing it even then.

But even just being able to do graphics, build small databases, build programs, and control our own destiny a little bit seemed fantastic. In that era, it seemed like there were ideas everywhere and we were all running after them. I can remember hearing about early versions of things like Photoshop and thinking "Wow, that's a cool idea," over and over. The old guys, the skeptical guys, were just irrelevant, and we raced on ahead and let them try to keep up with us. Some did, some didn't. I started running and, more or less, have never stopped.

Today, I see the same kind of opportunity everywhere I look. Not exactly like the PC, of course, but following the same pattern: There are lots of technical advances everywhere that can fundamentally change how we think about problems, just like the PC was a new capability that let us rethink engineering (and social and financial) constraints around all kinds of businesses. I spoke recently to the CTO of a large tech company who said, "Nothing has been exciting since the internet, until now. Everything is exciting again!" I completely agree with this. I don't know where to put my attention sometimes because there are so many interesting things to think about, across so many intersecting domains: biology, design, AR/VR, renewables, materials, and, of course, AI connecting all of it.

But, for whatever reason, the younger generation these days isn't kicking my butt the way I feel I did with older generations when I was their age. It's weird. I talk to students all the time now, and they are, almost always, very pessimistic. Not only do they not come to me with interesting new ideas and challenges, but they often don't think that things are that interesting, or that problems are that solvable. There is a lot of despair, but more interestingly, a lot of hesitation to even try. There are hard problems, sure, and some of them are going to be very challenging to solve, but somehow, many of the current generation (I would never say all!) don't feel capable to try, in what seems to me to be the most interesting technical environment in 40 years.

Let me give an example: climate change. The mood on campuses is almost literally doom these days: "We are

No Prize for Pessimism 5

the last generation," "Don't bother having kids," etc. I'll talk to groups and tell them things like "Solar and wind now beats coal economically and installations are exceeding all projections" or "It's now possible to make natural gas directly from the air, carbon free, at below market cost" or "There are numerous synthetic agriculture methods like precision fermentation that are thousands of times more efficient than what we do now, and more humane" or "We could easily afford now to build a solar plant that would desalinate enough water to fill the Columbia river — it wouldn't even be a particularly large check." And when I do, I mostly get blank stares, feeble pushback, or dismissal. It doesn't fit the narrative, so it's hard to think about! I've never once had a kid come up after with excitement, asking how they can help or who to talk to, or sharing a new idea sparked by some of this technology.

This is backward! The younger generation should be running circles around their elders. Sure, there are all kinds of things we don't understand, slang, whatever — but I rarely hear younger people say, "You don't understand, this is going to change everything." They don't seem to have the same kind of openness, experimentation, and ambition that I remember.

I'm not arguing that the generations after Gen X lack optimism and therefore fail to be innovative, but

I do believe the optimism-to-innovation quotient is diminishing among younger generations. In his book, *The Anxious Generation*, Jonathan Haidt writes that the Great Rewiring is a five-year period in which technological changes interacted with social trends to radically transform the daily lives of teenagers in the US and elsewhere. "In 2010, few teens had a smartphone, few had high-speed internet, few had unlimited data plans, nobody had Instagram and kids still sometimes went over to each other's houses and spent some time with other kids," Benjamin P. Russell wrote in his review of the book for *The New York Times*. By 2015, everything changed. Now they invest their time and effort in the virtual world, into things that don't pay off in the long run.

I'm a science and tech nerd. I am interested in all kinds of new things, and we'll talk about some of them in this book. But I'm also someone who likes to break things down into fundamentals and patterns (which is the point of this whole book!), so I started to think about why this was going on. The senior engineers I know (all the folks roughly my age or so) are more excited now than they've ever been, but the younger generation doesn't seem to be.

My experience is that it's a form of "detuning" that has to do with the information environment this gen-

No Prize for Pessimism   7

eration has grown up in. In my generation, we mostly interacted with "slow" things – the analog world, books on paper in a library, classrooms, single digit numbers of TV channels, print newspapers. We got a more or less normal distribution of events. Most were mundane, a few were sensational. Most TV was "meh," while a few shows were amazing. Most movies were forgettable, a few were Star Wars.

But that's not how the environment is for this generation. The same technologies that I like so much — the PC, smartphone, gaming console, augmented reality headset — are presenting a very distorted statistical picture now, and I think it's confusing this generation. It may get in the way of their ability to innovate and disrupt.

The simple version of this idea is that rare events are being regularly presented as "normal" in the statistical sense. On Instagram or TikTok, everyone's business is an instant success; everyone has had a great life; every idea is brilliant and there are never any problems or challenges. Or the opposite: Every disaster is horrible; every failure is a global humiliation that ends a career or gets you canceled; every risk is very high cost and to be avoided. If you're in this landscape, the message you get is that you have to be perfect at first, or you better

8   Sam Schillace

not even try. Experiments and failure, god forbid, are shameful and to be avoided.

We all know entrepreneurship, and building products, isn't like this. Well, actually "we" don't all know it, but anyone who has actually been through any process of creation — of a business, of a product, of art — knows that it's very messy and that there are always many false starts and setbacks. I was fortunate enough to have a role in building Google Docs, and that looks like a great idea and instant success from here, but as we'll discuss in the next chapter, that path is only clear in retrospect. In the moment, it was a thicket, with lots of doubt.

We all set our models of the world from what we observe. This is a fundamental part of being human. If you grow up in a low trust environment, it's hard to be trusting. If you grow up in an abundant one, it's hard to be frugal. And so on.

This generation is growing up in an environment that can be distorted. Everything is tuned for attention and impact, sometimes by algorithms explicitly, and sometimes accidentally by the nature of the tech, but tuned nonetheless. It's not the fault of people growing up in that environment that they are forming mental models that are inaccurate — it's exactly what we'd expect. But it's not really what we, or they, want.

No Prize for Pessimism   9

Another related thing that has happened with a lot of this generation is that they are groomed and managed for success from an early age. This has always gone on, I'm sure, but as colleges have been harder to get into, there is a segment of this generation that has essentially been project managed to never fail from day one. Their lives are curated and planned: Get good grades, do extracurriculars and lessons, be heavily scheduled, think carefully about college plans, don't just mess around with your friends outside and waste time. This training and mindset make it easier to get into an elite school, but makes it much harder for them to take the kinds of necessary risks and leaps of faith that disruptive innovation requires.

I spent a lot of time in my childhood just messing around, sometimes on my own, sometimes with friends. I grew up in a wild bit of land that had been given to a university (that my father worked at) by a car company founder. We had fields and forests and swamps to hang out in, and my friends and I would do all kinds of "dangerous" or messy things – making fires, building forts, hunting for turtles in a boat on a lake (imagine a couple of 13-year-olds doing that totally on their own today, no adults even aware it was happening). We explored every inch of that land and knew it well, played with anything we could find, learned how it all worked.

I got into a good school — the University of Michigan — somewhat accidentally (I think it's considered to be a more elite school today than it was when I got in). I was a good enough student, but only a 3.7 GPA, which wouldn't make it today. I was a math fiend, but only because I was such a science nerd, not because it seemed like a good thing to do for school. I took and passed a bunch of AP tests not because it would look good but because I didn't want to be bored doing the classes again. U of M was kind of the "default" for smart nerds in Michigan, where I grew up. I never did sports and barely did extramural stuff. Mostly I liked to hang out in the woods or my dad's "shop" in the garage after school, making things. Usually, things that didn't make sense to anyone but me.

But my kids were very different. We are pretty active parents, and both of my kids like the outdoors, but neither got to go wander around a 100-acre swamp as teenagers, and both of them thought about how to shape their resume to get into a good school. Being a teenager for them is less about an era of exploration and more a job in itself. And all of this plays out in public now — every triumph of your peers is amplified and your failures broadcast. The day kids hear about acceptance to high school (not even college!) is agonizing.

A hole in your resume?

Let's contrast that with something that came up in some mentoring I've done, which illustrates this point of the new generation being more curated and focused on failure.

I am asked to do mentoring often. One pattern I notice is that this "curated life" carries over into work now. Most people are focused on "How do I get to where you are?" which is natural, and they want a precise recipe: do this, get that. But when I tell them there is no set path or next step, they get very unhappy. Everyone seems to want to optimize for getting to the next level in their career ladder. I think that kind of "linear optimization" is fine, but it gets linear results. I never worried about this — I just did what I wanted to do.

In fact, this is the career advice I always give instead: Career success comes from impact. Impact comes from doing something you are good at that is valuable with as much energy as possible. We often have blind spots about what it is that we are good at; if it feels natural and easy, we think it must not be valuable (blame the Calvinists for the idea that you can only have value through suffering. Not so!) But the best possible career is when you feel guilty getting paid because what you do feels like play. And play is something this generation struggles with.

My first job in 1989 was for a company called Ashton-Tate. It was a big database company at the time, and my parents were happy that I'd managed to land a "good" corporate job. I moved out to California and began my "career."

The company began to struggle and ultimately failed about 18 months after I started; I left right before it finally crashed. My friend and I had seen this coming and had been writing a video game called Spectre as a distraction in our spare time. It was a simple game, but we put a lot into it. It was fast (which was very hard on the machines of the day — remember, underpowered PCs), multiplayer, and "twitchy," which we liked (there are some lessons from Spectre that will show up later).

When the company fell apart, we found a publisher for our game, somewhat accidentally. (A friend said: "Wow, this is great! Can I introduce you to someone to sell it?" We didn't know any better, and we'd really only made the game for ourselves, so … sure?) The game was sold and was moderately successful, we started to get some royalties, and my friend and I went off to formally found a startup together.

When we did that, though, everyone around us was worried — what happens if the startup fails? What would my resume look like with a "hole" in it?

Shouldn't I try to get another big company job, work on my career? I should plan and pay attention, not just go experiment! (As it turned out, that startup DID fail, but from its ruins rose another startup that paid for my house … you never can tell).

Fast forward 15 years, when I'm at Stanford's StartX mentoring new companies. I would hear repeatedly from founders something along the lines of "I need to do this startup so I have it on my resume." Now a startup, failed or not, was an asset, almost a must-have, instead of a liability. A complete inversion!

This is the beginning of that generational effect, I think. My generation didn't really think that much about whether a startup was a good idea or not; we just built things, and it worked or didn't. But the later generation, coming from a much more curated and careful background, thought of startups as stepping stones in a career, and were much more focused on whether they worked or not. It's not their fault; this is the environment they were "trained" in.

And many of them also struggled with these startups — doing a startup because you are "supposed to" is a recipe for unhappiness. Startups are always hard (getting the theme yet?), so if you are doing them for any reason other than really wanting to solve someone's problem, you will be much less capable mentally and emotionally of getting through the tough times. The mindset of abundance and experimentation that we will explore in this book is really important here.

The challenge with planning a startup as part of your resume is that this isn't how disruptive innovation and problem solving really works. Innovation and disruption are messy by *definition*. They are almost impossible to plan fully, and they require a tolerance for experiments, failure, going backward, and putting up with slow progress sometimes.

In business, and science, if the path is obvious, it's usually already been taken. Easy and valuable things are usually crowded and done quickly. This is almost definitional too: It's the height of arrogance to think that there might be an idea that is both good and obvious, but that somehow no one but you has noticed it. It's not impossible — sometimes that does happen! But it's very rare. The more usual case is that things are less obvious and riskier. Risk begets reward, right? The real value is almost always in directions where people are skeptical, the science is a little shaky still, and the solution isn't really fully clear. You might fail! In the next chapter, we will dig more deeply into this in a real-world example, the early days of Writely and Google docs.

No Prize for Pessimism   15

We're going to explore a lot more of these patterns in this book, and hopefully find some good tools that everyone can use to navigate this complex world we find ourselves in. In some ways, the world itself is akin to a technical organization or large codebase now. There are lots of moving parts, unexpected interactions, complex systems, and incentives — and it's constantly changing. Some of the observations and tools from my time as an engineer and leader are directly applicable not just to inventors and entrepreneurs, but to anyone who has to navigate and understand that complexity.

The good news is that it's achievable. Complex systems move in relatively predictable ways, and they have patterns that reappear reliably. Innovation, too, has principles, best practices and mental mindsets that can help make you more successful — though it is always a tough process to hack through the "thicket of the unknown." Each innovation is different and surprising, each engineering challenge has to be approached anew, but they're all the same at some fundamental level. We make progress by breaking large problems down into smaller ones, working methodically, looking at scale bottlenecks, and asking "What if?"

There are many challenges in the world that do need to be solved: climate change, financial inequalities, social

16   Sam Schillace

and political challenges. And we are getting more capable all the time, which is as much dangerous as it is good. We need younger generations to be engaged in these problems in a meaningful way.

One main dichotomy that I see in my speaking and working, is the divide between optimism and pessimism. This shows up in many places and it worries me. One is in the contrast between an "abundance" and an "austerity" mindset. I think the new generation is much more in an "austerity" mindset. Unfortunately, that mindset usually leads to less interesting lives in the near term, and social unrest in the long term. When we've been in that mindset, we've caused economic harm, wars, and dictatorships.

Thomas Robert Malthus was an English economist and demographer who lived from 1766 to 1834. He is best known for his work *"An Essay on the Principle of Population,"* first published in 1798. Malthus predicted that population growth would outpace agricultural production, leading to widespread famine and societal collapse. He argued that while populations tend to grow geometrically, food supply only increases arithmetically, creating a gap that would result in catastrophic shortages. Malthus had the ultimate austerity mindset — the idea that we would eventually just run out of everything.

So far, none of the Malthusian predictions have come true. Why is that?

Well, it's because humans are fundamentally clever and create new value — the "abundance" mindset. Until about the 1850s, it was thought to be impossible to grow much past where we were. Most people were farming most of the time, sustenance was hard work, and we could only grow so much. Then a brilliant German engineer figured out how to make "bread from air" — fixing nitrogen from the atmosphere. Suddenly, because of that human cleverness, we had abundance: much more food. And that dovetailed nicely with the industrial revolution, letting people get out of the business of manually farming, spreading fertilizer by hand, etc.

About 2 percent of the population is now involved in growing food. Another inversion, this time a good one! There was disruption and unrest as we went through that transition, for sure (and that tech helped lead to two World Wars, sadly), but now, I doubt many of us would like to go back to the world of "no tech, no computers, no medicine, just work on the farm until you're exhausted for your daily bread." It's a kind of austerity that might sound good on paper, but is in fact a gigantic human crisis if we adopted it. Humans are awesome — we want more of them, and we want them all to be happy and creative! That's abundance.

And, interestingly, we might not be done yet with even something as fundamental as agriculture. Sure, there are some not awesome farming practices like selective herbicides that have bad side effects, but there are also all kinds of farmers that are using modern tech to be better. I've talked to farmers who are using AI to understand what is going on, meter by meter, in their fields, and automatically adjust fertilizer and water to optimize their crops. There are lots of ways to make small farmers more effective, and we all (at least here in the United States, for the most part) have access to much higher quality, lower cost, and more diverse farm products than we have ever had.

But beyond that, there are even more exotic things coming like "precision fermentation." There's a company, Perfect Day, that makes the milk protein casein from engineered yeast. It's much more effective than using cows (which emit methane anyway). How much more? 100 times? 1,000 times? Nope: 3,000 times. Same protein, safe, used already in ice cream and other food products. What else can we do that with?

Okay, at this point if you are a good "glass is half-empty" person, you'll be thinking "yeah but that will be abused. They'll make addictive milk or something." It might be. The biggest risk in new tech is almost always people, not the tech itself. And this is another reason why the generational "inversion" is so urgent to understand and deal with.

We have to remember that people will take whatever solution seems easiest. If you want there to be a solution to a problem, like producing casein, that is both better and good for the world then you have to achieve that by creating it in a way that's easier to consume. Just wishing that someone wouldn't implement an obviously abusive, but lucrative, business isn't enough. You have to beat them somehow in the marketplace of ideas.

When we built GDocs, it wasn't because we wanted to "hurt" Microsoft Word or had some moral qualm or whatever. It was because we wanted there to be something better for our specific tasks. So we built that, and we largely ignored anyone who didn't understand it, until mostly everyone did.

Generations are, however, generational.

◆◆◆

The American writer Gertrude Stein famously labeled her generation, those who fought in World War I and came of age during the Roaring '20s, the Lost Generation. The writers and artists of this generation gathered in Parisian salons and cafes to discuss and debate ideas and imaginative creations that would influence the world. Another generation, the Baby Boomers, came along just after World War II. Boomers and the GenXers who followed them comprise the core of senior engineering leadership in tech today.

For my generation, the Parisian cafes look more like urban coffee shops, college campuses, and startups. We're no smarter or capable than the generations of programmers and developers who have entered tech from younger generations, but, as a whole, we seem to be more optimistic about what's possible, and as a result, we are often more dogged in our pursuit of innovations we wish to see.

But we're aging. We've lost many titans, including Steve Jobs, and others are approaching retirement or at least scaling back. What is our succession plan? Whither innovation?

As a tenured leader, I've certainly noticed this alarming trend of pessimism and doubt among some younger colleagues. While the gray-bearded engineer works to pump up the team with enthusiasm, too often the younger engineers worry the idea won't work. Granted, this can be for good reason. Their lives have consisted of terrorist attacks, great recessions, climate change, a global pandemic, growing mental health concerns, the comparison culture exacerbated by social media, and, yes, the technological uncertainty of AI.

For many in my generation, we were shaped by the idea that we could send a man to the moon and return him safely to the earth — all within a single decade. Later we watched in amazement as a vaccine against COVID-19 was created, tested, manufactured, and distributed to anyone on the planet who wanted it in a matter of months.

What these miracles of innovation have in common is smart people, but also people with the optimism and determination to succeed. Where does this optimism come from? Are you just born with it, unlucky if you don't get that gene? In the rest of the book I will argue that optimism, invention, and disruption are all choices — a mindset and approach we can learn, practice, and get better at. There's no prize for being pessimistic and right. Let's go learn how to disrupt and invent!

Chapter 2

Google Docs:
A Story of Invention

In my personal Google account lives the very first Google Doc. I shared it with my cofounder Steve Newman in late 2005 when we were working on the product.

Today, I don't know the exact number of documents created or even the number of users, but it's well above a billion people who use it every month, and it's possible now that it's over 2 billion. So, from where we sit, it must have been a really good idea, right? Uncontroversial, obvious, and easy to do.

Not even close.

What looks like a clear, obvious path now was actually a confusing, difficult thicket of challenges at the time. In this chapter we are going to look at some of the lessons learned from that history, and some general patterns we can derive from them.

When I say "the first Google Doc," that seems like a relatively simple concept. I made a document. It still exists, therefore, it's *the* first Google Doc. But it is more complex than it seems, and the history is more tangled than you'd expect.

In the summer of 2004, a company I founded with Steve Newman and Claudia Carpenter was trying to find a reason to exist. Steve and I had founded companies before, and we had met Claudia at Intuit, where we were hired to "help teach Intuit how to be more like a startup" by building some software. The three of us had left in 2003 and had spent time wandering around looking at technical ideas. The current thing we were working on was some kind of distributed bug database; who knows, maybe we would have figured out GitHub if we'd gotten lucky!

Steve and I were in the habit of exploring technical spaces that intersected with our interests, but this idea didn't seem to have much traction. We'd built an experimental database that was a little easier to use for web applications. You have to remember, in this period, the term "cloud" wasn't widely used, and the tools for building online applications were still nascent. Gmail was the only real website that could be called a dynamic application, and all of the best practices and programming techniques were still being invented. We were actually using .NET and C#, which the first version of GDocs was written in. (As a fun aside: I often have people saying "thank you" when I tell them about my history with GDocs. I recently met one of the people who created .NET, so I got to say, "Thank you for letting me build something that people thank me for," which was fun. The valley is kind of a small place, it's very satisfying when you cross paths like that.)

That database ran on three Windows machines we had rented in a datacenter in Texas. That was all you could do then, if you were small: rent whole machines and manage them yourself. We had a sys admin who was in, I believe, the Philippines, who helped with them. I wonder if that person even knows that they helped with GDocs in the early days! I don't think I ever even met them.

That database is the first of three (that I know of) places that document has existed. Later on, it was moved into Bigtable, and then after that to Spanner, at Google (and we will talk about some of those challenges in a bit). So the bits have moved around, and even been reformatted into a different file format.

But not only has the storage and format of the document changed, the code that displays it, called the "frontend" has changed at least twice too! Once when we moved from C# to Java, and then again when it was rewritten at Google to perform better. I'm getting ahead of myself a bit and we'll come back to that part of the story, but the point is that even something as simple in retrospect as "this is the first Google Doc" is subtle and complicated. It's still "logically" the first document in some sense, but no aspect of it — not the contents, the storage format or the rendering — are exactly the same as they were in 2005. You could almost argue that an NFT of "the first Google Doc" would be more durable and representative than the document itself!

Why does any of this matter? Well, what often happens in innovation and disruption is that things are "wrong" until they're "obvious." We forget that there have been many changes and evolutions along the way. It's important to remember that with all technology, it's a continuous dialog. What seems like it always existed, hasn't, and will stop being what it is now at some point. Learning to notice this as it changes is hard, but critical. It's very easy to get caught up in the moment of some story — this is how things are NOW, so they have always been and will always be. But things aren't points; they are lines and curves. Being an innovator means being aware of this and learning to think in terms of how things might change and evolve over time.

26   Sam Schillace

Back to the history of GDocs. In the summer of 2005, as this startup was trying to find its way, I stumbled across two different new browser capabilities. I don't think they were super new to the world, but they were new to me and becoming more widespread in the browsers that users had. The capabilities were JavaScript, and what was known as content-editable (sometimes written as one word).

Let's take content-editable first. The job of a browser is mostly to display the content of a web page, which is created and stored in a format we all know: HTML. At that time, browsers mostly just displayed "static" HTML. A program running on a web server somewhere would generate a page by creating some HTML, and then send it to the browser, which would "render" it as a web page. If you wanted something else to appear, you had to (for the most part) ask the server to create an entire new page, send it down, and have the browser redraw all of it.

Content-editable let you put the browser into a mode where part of the page was editable without having to go back to the server. You could just edit directly in the browser. This sounds pretty good but at the time it was very incomplete; okay, you've edited some of this HTML that got sent down to the browser, but now what do you do with it? The browsers don't really store HTML for long periods of time, like a file system does; they throw the web page away as soon as a new version comes down.

One thing we saw initially with content-editable was that lots of teams sprang up to do essentially the same thing: edit some content and then save it ... somewhere? That tended to be left up to the user. You could, I guess, integrate with an existing workflow or file server. But then you also had to get the new content back into your web server and get it served out again. It was complicated and hard.

This actually illustrates a fundamental principle: Users are lazy. And that's not a criticism; it's the fuel for our creativity and innovation. It's what drives us to continue to iterate until we've made these products accessible to everyone. In the case of the many small companies that just provided a thin wrapper around content-editable, they didn't really make their end users' lives much better, and so they weren't of much value, and didn't go anywhere. Though we didn't fully understand it at the time, we took a path that let users be very "lazy" and get their job done quickly and easily. There were many parts of that solution, but an important one was taking content-editable, and not just presenting it as-is, but looking at what it could do: how we could use it to edit something that felt more like a document, and what that would mean.

But that was a bit later. The second thing I noticed that summer was JavaScript. That had been around for some time, but it was starting to get fairly useful and was, at the time, in something like 85% of the installed browsers (This is an important number: After we got the first version of Writely, the product that became GDocs, working, I had to go check to see if we even had a market we could get to).

JavaScript was controversial at the time. Remember the "old guys" in the first chapter? I was becoming one of them. Programming languages until then had a few consistent features. One was that they were "compiled." This means that you run the language through a program, called a compiler, that turns it directly into machine code, which is very fast to run. JavaScript was "interpreted," which means that a different kind of program reads the programming code as it's being run and executes instructions on a kind of virtual computer known as a "Virtual Machine." This is faster for the programmer — you don't have to wait for the compiler — and it's easier to move around, since you don't have to build different machine code for each machine, you just have to have the VM there. But at the time this wasn't considered "production ready."

JavaScript is also "weakly typed." This has to do with that idea of precision we mentioned in the first chapter. Programs, until then, were fairly strict about what kind (or "type") of object everything is. A number could be an Int, which is like a whole number, or a Short, which is a whole number less than a certain value, or a float, which is a number with a decimal point, etc. In JavaScript, there were no types. You just had numbers, and strings, and sometimes they changed back and forth. Again, this wasn't considered production ready. But because the code ran on a VM instead of having to be compiled, the system could look at what each object was as the program ran; it made it much more flexible and easier to use.

JavaScript also had some properties that made it well suited to working in the browser. One was that it's what's known as a "functional" language. Most programming languages are like a story that is told in a linear thread, start to finish. There is some jumping around, but there is only one place where you "pay attention." In a functional language, you can "pinch off" a little bit of code, and have it be called later, out of order. This is useful, for example, if you want to make a call to a server that is far away and might take a while, and you want to do things for the user in the meantime, but then go back to that context and take some action when the server does respond. We used this for doing things with editing; I could send a change to the server, and then "pinch off" a function that would integrate the server's response when it came back, and then let the user keep editing while we waited. If I had to wait for the server — to "block" in programmer parlance — then editing would become very awkward and hard.

So those two pieces — content-editable and JavaScript — came to my attention in the summer of 2005. Steve and I had both worked on word processors in the past. Way back in the '80s Steve had built something called FullWrite, which was one of the first WYSIWYG (what you see is what you get — that is, fully graphical) Word processors; we had both created Claris Homepage, which was a web page editor; and we both worked on Dreamweaver MX.

So it was natural to look at these two capabilities and think "Let's try to build some kind of word processor in the browser." And that's it, it worked on the first try, no controversy, end of story.

Of course not. In fact, even the idea wasn't obvious. Mostly what all of that experience with word processors got us was skepticism. Both Steve and Claudia (who also had a lot of experience with application building) said, "There is no way the browser will support a satisfying word processing experience; let's not do it."

Here's the interesting thing about that. They were right ... and wrong ... at the same time. The browsers of the day absolutely couldn't support what GDocs has become now. Remember that word, "Virtual Machine"? The VMs of that era were very limited — they were slow; not much effort had been put into making the code they ran as fast as compiled code, which it almost is now; and they didn't let web pages run very much code at once. Nothing was standard across the different lines of browsers, and there were bugs everywhere. It was a mess!

But it was also the case that as soon as the cloud emerged as a new paradigm, as soon as the tech industry saw how powerful the new model was and how useful the apps could be, there was a tremendous "pull" to add functionality and improvements to make it work. People wanted it to be the case that there were rich applications in the browser, and the early version of Google Docs helped motivate this.

But at the time, we didn't know what GDocs would be, or that the world would help so much. So how did we overcome this skepticism? Well, to some degree we didn't have anything better to do. But more importantly, it was a cheap experiment.

Cheap experiments are the best. Usually when something is disruptive and new, you're skeptical about it. Why is that?

I think this has to do with defending ourselves — both physically and emotionally. The physical is easy. In our archaic environment, new things were probably more likely to hurt than help you, so caution is advised. That's probably baked into us at some low level. But in our nice, safe, comfortable modern lives, we are still skeptical of new things inherently. Why? I think there are a few things going on. One is just basic laziness; we don't really like to work hard at thinking. New things force us to do that. It's work! Skepticism is a good story to tell yourself about why you don't have to do that work. "Well, it's probably dumb anyway, no sense getting up." And maybe that's usually true, but when there is something interesting to learn, you won't learn it if you are too skeptical to try.

The other reason. I think, is ego protection. New things challenge our sense of ourselves and the world. That's painful; it can mean we are wrong, we lose status, we have to do more work to react, etc. It's far easier to tell ourselves the story that the new thing is likely to be wrong and feel good.

We can examine our reasons for skepticism, and we can find ways to attack our natural tendency to reject new ideas — but one of the best tools to overcome our skepticism is to just try things cheaply and quickly.

In our case, we had the database already, and we didn't have to do much work: The browser would supply the editing surface for us, and I could write the code in JavaScript. No install, not much to learn, just try it with tools we already had, so we did. The first version took about a week to build.

It worked … okay. We were already a "distributed" team that often worked from home and came together a few times a week. In the tech of the day, there weren't many good ways for a team like that to share and manage documents. You could send email attachments, but then you'd always be out of date or have the wrong version — plus that's work! (users are lazy!) You could set up some kind of file server but that had other problems: Sometimes the files would be "locked" when one person was working on them (and, what was worse, sometimes people would forget to unlock them and then go home for the weekend), you had to install desktop software and mount the shared drive, which was fiddly and hard (user are lazy!), and you had to make sure your file server

34   Sam Schillace

was visible to you and only you, configuring a firewall securely (also hard; users are lazy!).

Shared documents, in the browser, with no installation, and simple password-based authentication (which we didn't even bother with at first) were AWESOME! It was very apparent to us almost immediately that there was something here. And then ... we started to collide with each other's changes, almost immediately.

The cycle began. For the next few years, we would work on something in GDocs until we broke it or found something that didn't work, and then we'd iterate until it was right again. This happened first with collaborative editing, but it continued with all kinds of things: importing documents, dealing with Google storage infrastructure, setting permissions, document organization, versioning, and lots of features.

But for now, we needed to be able to edit without losing changes. Our first, naive, implementation was what was called "last in wins": whichever edit got to the server last would just overwrite whatever was there. It was crude, and we lost edits. We needed something better, so we decided to do something called "three-way merge."

This is a merge strategy that has since fallen out of favor (we have something called "Operational Transform" now which is much more reliable). The idea is that you have two changes, from two different users, that you need to reconcile. Those are two "versions" of the document. There is a third version, the reference version, that you built the last time a user made a change that was accepted by the server. Each of the two user versions can be thought of as a change relative to that reference, and they can be merged into a new reference version: the three-way merge. Then a new reference version is created, and changes relative to that reference version are sent back down to each user, so they can update their document to import changes from the other user. Makes sense, relatively simple … and it doesn't really work.

Why doesn't it work? This was the first of many engineering challenges. Remember that HTML format I mentioned? It turns out that there are many different ways to create the same document; you can think of them as dialects or accents maybe: different elements of the HTML can be in different orders, have different capitalization, even different counts.

What that means in practice is that, if you're not very careful, there can be "browser fights." One user makes a change. We reconcile it, send it to the other user. That user's browser takes it but transforms the HTML into its dialect. This looks like a change, so that gets sent back

36   Sam Schillace

to the server, which reconciles and sends it to the original user, whose browser does the same kind of change and sends it back again … and around we go.

I won't go into too many more of the technical details here. Think for a second though, about that early skepticism. If we hadn't seen how valuable the editing model was, and instead had noticed the "browser fight" problem first, do you think we would have gone on to even try the experiment in the first place? I don't know that we would have! That problem seems hard; if you don't know that it's worth solving, would you try?

All innovations, particularly disruptive ones, have this pattern to them. Obvious, easy-to-implement ideas are usually … implemented! It's very rare to get to one first because whoever did the step before will usually have seen that and done it too. So we are always confronted with this kind of implementation challenge, no matter what.

This is one of the most important points in this book. I think of it as asking "What if?" instead of "Why try?" It's really easy to find — even to make up — reasons not to try something. Particularly when something is really novel and challenges our world view, there is a real tendency to feel defensive first and then to find a reason for that feeling. "Merging is going to be really impossi-

No Prize for Pessimism 37

ble to get right" is a really good example of "Why try?" (and we'll discuss a whole bunch more of these that were asked about GDocs over the next few years soon).

"What if?" questions ask "What if this works? Is that cool?" In our case, we were extremely fortunate to know the answer: We already knew it was kind of awesome to do this kind of editing and collaboration. So when the "Why try" problems started to appear — and they did — we were ready for them. One of the most important things you can do as an entrepreneur is focus on your own "What ifs." The world will find plenty of problems for you to solve, but only you can have that vision and keep it as your north star. And the example of Writely is helpful too: Cheap experiments are a good way to develop conviction on your "What if" questions. If you've seen it working, it's much easier to imagine the future.

This breakdown — why try versus what if — is at the root of a phenomenon that mystified me at the time. We built out Writely into a more capable program and started advertising it. We had in mind that we wanted to be some kind of subscription service, like BaseCamp from 37Signals or something, and we were trying to see if we could acquire customers in a cost-effective way.

38   Sam Schillace

But as people began to encounter it, we noticed that there was almost no one who didn't have very strong feelings about it. A small number of people were incredibly enthused about it (shout out to Nate Torkington — love you for your support, man!), sometimes more enthused than we were, so enthused we were even confused about it. But a lot of people HATED it at first.

One day I received an email from Google asking if we could talk. I went to the Google campus, and one of their corporate development guys met with me. He looked across the table and asked, "So, it's just you?" I told him there were three of us and he seemed very unimpressed. It became clear Google was not going to offer us much money.

Another tech giant in the valley also became interested and word got back to Google's Eric Schmidt, who I'm told banged his fist on the table and screamed that he wanted the deal to be done to acquire Writely. And the deal got done. We went to Google, and Writely became Google Docs.

At this point you may be thinking to yourself, I'd be optimistic, too, if Google bought my little startup. Not so fast! We were *not* on a magic carpet ride. For starters, Google had never worked on software in which users could make changes on their own and see the change happen instantly on the screen. Our product enabled a user to edit and collaborate online, live. Gmail was the closest, but Gmail was sequential, meaning the tool just stacks up changes very simply. So it's serializable. But our product enabled multiple users to edit all at once. Nothing Google had built was capable of dealing with that.

To succeed with expectations, we needed a lot of hardware, but our new manager simply didn't get it and refused to get us what we needed. I walked into his office and said we can either go talk to the boss, Eric Schmidt, together or we can sue you. We got our hardware.

A young Marissa Mayer, employee number 20 at Google and the future CEO of Yahoo, saw our product as a rival to one she had been championing. That project was called Page Creator, and they wanted to suck us into their effort. We declined and held fast to our vision for Google Docs.

Next, we had to migrate our product into the company's datacenter, which was challenging. We had to learn 12 different internal Google technologies in just three months, but we did it. We migrated our code from C# to Java, which was a difficult process. You can't imagine how hard it is to do migration on this scale under these time pressures. You think it's a good day when there's only 150,000 compiler errors remaining, and you think you'll be done by the end of the week.

We somehow mastered Google's internal authentication system, storage system, networking system, message delivery system, and many other systems I can't even remember anymore

Through it all, we managed to balance being stubborn and persistent with being open-minded and flexible in the face of internal complexities and market feedback. We managed to launch faster than any acquired company before us.

At any point it could have been easy to become pessimistic or cynical about innovation. There were many low points, but I knew we were doing something important. Today one-quarter of the world's population uses that product. It has revolutionized the way people collaborate on documents. It's made the world more productive. It's advanced education, reduced impact on the environment, and spurred innovation.

Not every idea will turn out to be a Google Docs. Most of them won't. You have to be careful (especially in the current era of very selective social media) of things like selection bias. No one ever shows you all the bad ideas along the way, but trust me, we all have them.

The idea that preceded GDocs was a real stinker (peer-to-peer bug database).

The only strategy is the one we are discussing here. Do lots of interesting things, as fast and cheaply as you can, learn a lot from them, discard them or iterate quickly, and just keep going. Nothing really changes the ratio of good ideas to bad, but you can try a lot more ideas, you can raise the quality somewhat and you can learn to learn better and faster. All of those together make it much more likely you'll find something disruptive and valuable.

Letters from a Messy Tech Optimist

Chapter 3

Mindset: Approaching Disruption and Innovation

When I started writing my *Sunday Letters*, and even now as I sit and write this book, I imagine the reader to be my younger self (or a young engineer just starting their career). My younger self was skilled and hopeful but not yet fully optimistic, and not yet willing to never take no for an answer.

In this essay and those that follow, I have curated a selection of *Sunday Letters* from over the years. Programmers of a certain age will remember Jon Bentley's classic book, *Programming Pearls*. "Just as natural pearls grow from grains of sand that irritate oysters, programming pearls have grown from real problems that have irritated real programmers," according to the book's overview.

It is with that spirit that I offer these essays. My aim for the reader is to curate and share the spirit of optimism and innovation I've experienced during a span of computing revolutions, from the PC to the Internet, cloud, mobile, and now AI. These essays are meant to be a combination of inspiration and information (at least I hope they are!).

All advances are by breaking the rules

One subject that comes up a lot in these letters and in my conversations with folks is the idea of disruption, and "disruptive innovation." I am a fan, and always have been. But why is that? Why is any kind of disruption, experimentation, willingness to fail, etc., considered "good" at all? Why don't we just stop and enjoy what we have, or go back to something else? Why do we encourage risks, fund it with venture capital, and celebrate ambitious founders?

There are lots of opinions here, and lots of folks talk about reasons to be optimistic (or pessimistic) about technology and change in general. It's certainly true that not all change is good, and there are better and worse ways to approach disruption. But why support it at all?

For me, the answer is very urgent, and very fundamental — and it's contained in the title of this post. Every single thing that we have invented — starting with fire and moving up to <pick your tech of the day> — was "not the way we do it" until it was done for the first time. Maybe there wasn't an explicit rule against it (though usually there was), but there was at least orthodoxy, perhaps dogma, and often social pressure.

For many of these changes, there was no permission, and often extreme social pressure against them (pace Galileo). This will always be the case; there will always be those who can't or don't want to understand new things. Sometimes they're in positions of power and they can block the new ideas. It's been said that science advances "one funeral at a time" for this reason. We have to wait for the senior scientists who adhere to old dogma to retire or die off before new paradigms can be established.

Nothing improves without changes and experiments, and there are no experiments without failures. One way of looking at evolution is that "life is 4 billion years of mistakes." Every genetic change that has "advanced" life on Earth was a mistake of some kind in the transcription of genetic code. We kept the ones that were happy mistakes (fitter), and the unhappy ones didn't survive.

So there is almost a moral dimension or imperative to invention. We are almost obliged to poke at the status quo, ask questions, try new things, challenge "the way things are done." I am very far from being a political scientist, but this seems to me to also be a big part of why freer and more open societies tend to move faster and create more wealth: If there is a single authority saying what can and can't be done, you'll never take the risks and make the mistakes you need to improve. No one is always right, and this goes for managers, too. If you don't build a culture where you can be challenged as a leader, you'll leave a lot of value on the table.

There is a line, for sure. Breaking rules doesn't mean breaking laws or causing active harm. We are aiming for positive sum, not zero or negative. But we can't get better without trying new things. Disruption, innovation, experimentation, and failure are as important to us as breathing — and as fundamental.

From error, virtue

I make a lot of random things as a hobby. I am not particularly coordinated or skilled, so I make many small mistakes. I also work with unusual materials or ideas. As one example, I took some very old redwood that had been discarded from a logging operation 100 years ago (so it's probably old growth, but it's been sitting on the floor of a forest all that time, so it's pretty but delicate) and made a box out of some of it. But I somehow misaligned the cut — hard to know how since the entire thing is symmetrical– so that there was an even offset to the lid on both sides. It still fit, but the grain didn't align. So I took some nice Washi paper and covered just the small part of the misaligned grain, and then added some of the paper to the inside of the box and bottom of the lid. Not at all the design I started out with, but a really cool design at the end of the process.

As I've said before, one way to describe life is "4 billion years of mistakes." It's pretty accurate — the "goal" of reproduction is to reproduce, but sometimes there are mistakes in that process. Many of them are bad, and they die out. But some are more fit, and they thrive.

This is true in music, too. If you are jamming, and you play a "wrong" note, it sounds bad. If you stop, it's worse — but if you use it as a "passing tone" to a better note, it can sound great. The blues notes and those great bluegrass slides are "wrong" notes used the right way. There's a saying, "You're never more than a half-step from salvation" (I'll talk about this a bit more in another letter in this chapter).

One more story, this time from science. Stainless steel was discovered because an assistant put 1,000 times as much chromium as was called for into an experimental alloy. A few months later, they noticed that ingot, shiny in the rusted pile of rejects. There are many other stories of "mistakes" leading to insight.

I've decided this pattern needs to be called out and celebrated, and I've decided to call it "From Error, Virtue." I made a crest for it using the AI system DALL-E, and of course, the text is in error, which, to me makes it that much better:

We talk a lot about cheap experiments, being willing to fail, and things like that around here. We all try things that don't work and make mistakes even when we have the right idea. This can be your raw material for great things if you are flexible and aware. Keep your eyes open: There is virtue in error.

How do we make sense of all of this?
It's okay to be overwhelmed right now.

A lot is going on in the world right now (June 2023), especially with the explosion of AI news that's hitting all of us constantly. I spend a fair bit of my time reading and thinking about this as my day job, and I think I'm a pretty good pattern-matcher, and even *I'm* having trouble keeping up.

One big problem is that so much of what's happening has at least the potential to be very disruptive. The implications aren't incremental; they may have to do with how we work, create, communicate, transact, and more. There are hard questions being raised about what the nature of work is, what the role of intelligence is going to be in creating value, what harms and risks we need to pay attention to, and more.

There aren't going to be easy answers, at least for a while, and there are going to be false leads, early bad takes, all the usual "fog of war" stuff. I think there are three things we can look to right now that help navigate this.

The first thing is, fittingly, **first principles**. There are all kinds of patterns and behaviors that seem to show up repeatedly. Picking something you believe in and applying it to the current situation can help give clear framing. A few first principles that I like at this moment are:

- People are lazy. Look beyond "cool" to how much easier a new tool or tech makes someone's life. Convenience always wins.
- We reject new things by default, so try to ask "What if" as much as you can.
- We tend to be bad at understanding exponential curves, so we overestimate impact in the near term and underestimate it in the long term.
- All complex systems have scale constraints. When these are relaxed, the system reconfigures. Look at which constraints are being relaxed at this moment.

The last one leads me to the second tool we have: **math**. This is a good moment to set aside emotion and look at hard data where possible. I've seen people take things like the Chinchilla scaling paper and apply the math there in all kinds of useful ways that help us understand the costs of scaling and the likely curve of improvement. Math is a good cure for getting fooled by your linear intuition in a non-linear moment. Sometimes, you can combine the first two of these and apply some math to a first-principle idea (for example, people have been trying to quantify job efficiency improvements with AI).

Finally, the last thing is **an analogy**. Sometimes, we can look back at earlier technical transformations and draw parallels. This is the weakest of the three and has to be used carefully because it's easy to find false analogies. Analogy is more useful to find a starting point that you can then apply some more rigorous analysis to, per above. But sometimes, it's very useful to give a good sense of what's likely.

The one challenge with analogy is **dimensional reduction**. LLMs are incredibly complex and high-dimensional objects — they are like humans who have been reading for millennia. We don't have good mental models for something with that degree of complexity, and the behaviors fool us into thinking they are more like human minds than they really are. When we project these complex, high-dimensional objects down into a simpler analogy, we are doing dimensional reduction, which loses data and can be misleading if we don't do it carefully. Take it with a grain of salt if you can.

It can be challenging to keep up with everything that's going on, but it's not impossible. There are patterns and methods of analysis that can help us work through all of the change thoughtfully.

The prize for saying no is small
and who cares, anyway?

I've spent a lot of my life being very pessimistic for some personal reasons. I've written about this mindset before as the difference between doubting and asking "What if?" — I was a hardcore doubter, taking pride in myself when I was right.

This is one of the dumbest things I ever did in my career. For one thing, with very few exceptions (I'm looking at you, crypto), if there is enough value being pursued, they tend to get solved eventually even if there are early problems. There's really no point in being negative in this context. If you believe in the value being created, you should put that energy into solving problems, not criticizing. As others have said, the prize for being negative and right is very small.

And the penalty for being optimistic and wrong is also usually small, but it can feel uncomfortable in the moment. If you are an early believer (or at least a suspender-of-disbelief), you can feel foolish if the idea doesn't pan out. You were naive or you let your excitement get away with you and, with the benefit of hindsight, you can see the enthusiasm was misplaced.

That's okay! It's (hopefully) part of life to learn and grow, and to change our perspectives as we do. I constantly look back at earlier versions of myself and think "wow, I am SO much smarter now!" and then think "yes, but there is a future version of me out there, looking at me in this instant and shaking his head."

That early energy and enthusiasm, even if it turns out to be misplaced, is much better than caution, which is another way of saying no. Enthusiasm and suspension of disbelief are ways of asking "What if." Cynicism and caution are ways of saying, "It won't work." There are huge prizes for saying "What if" and being right. There are no prizes for any of the other choices. Play is rational!

In praise of small weirdos

Most of our experience as consumers of new technical products comes at later stages in their development. Just from a purely statistical perspective, unless you have a very active practice of seeking out new things, you'll be in a later and larger cohort when you encounter something for the first time. It's more unusual to encounter something in the early, messy stages.

This biases our understanding of how new products and tech come into existence. We already have a "narrative bias" that pushes us in the direction of telling neat stories about how things work. When we combine this with usually coming across things after some of the details have been worked out, this gives us the false perspective that, when trying to find innovative ideas, we are always on the hunt for ideas that are well-formed.

But the reality is that the valuable early signals are all messy and complicated. It's also the case that once you've spotted something that is well-established, it's usually on the late side, so it's harder to create new value and be competitive.

The right things to be on the lookout for are "small weirdos." Small means that things are still at the early stage — not many people "get" it or are using it. You haven't read about it in the press yet. "Weirdos" means that you initially might not understand why this small group of people are so enthusiastic about it, or, if you're really early, you might think something like, "That's weird, but I like using it."

All kinds of new tech fit this pattern early on. You can see this right now across a lot of AI-related tech; for example, there are interesting, weird things happening that aren't widely adopted yet.

It's hard to stay sensitive to this pattern. We like social proof and we like status quo, so "something weird a small number of people are doing" has a tough time getting through those filters. But if you want to be an innovator and stay at the front of developing tech, small weirdos are where it's at.

Asking the dumb questions
Beginner's mind

When you come into a new situation, it's natural to start out by being quiet and observing. Mostly, this is a good thing to do — you don't know what you don't know, so looking and listening is a good early strategy. You'll see things that don't make sense, but they will begin to make sense as you start to understand the whole picture. This is true for new teams, new domains of expertise, and new cities alike.

But sometimes, you'll see something that doesn't make sense, and you can't get it to make sense. It's tempting to just keep quiet, but you actually have a superpower here: You can ask the "dumb" question!

There's a word for this that is used in the study of Japanese martial arts: "beginner's mind." It's often the case that beginners are unexpectedly better at some aspects of a complex skill than they "should" be, and this is because they don't have all the preconceived ideas that the experts do — sometimes, they really can find more natural ways to do things. In fact, this is something that is actively cultivated as you gain experience — the goal is to simultaneously be a skilled expert but also constantly be seeing things through "fresh" eyes.

If you're new, no one expects you to know everything, so you can get away with asking something that seems "dumb" but actually isn't. In fact, you'd be surprised how often other members of the team have had the same question but are less able to ask it or to challenge something that seems settled. It's a very common pattern in teams: Sometimes things "just happen," and everyone assumes that everyone else knows why.

Often the new people can expose this in a healthy way. And if someone on your team asks something and you find yourself answering some variant of "just

No Prize for Pessimism   57

because," it's a good indication that there's something not right that deserves to be re-examined.

Trying to cultivate a beginner's mind is a very valuable practice. Giving any team member, but especially new ones, space to ask awkward questions is one of the most effective ways to keep your team nimble and growing. Celebrate those dumb questions! They're incredibly valuable.

Learning to learn, playing slowly
Embracing the slow and messy

It's easy to get into a mode of procrastination when confronted with learning something new. The bar is high, other people know better than you, you don't know where to start, and your early efforts are unsatisfactory. So you put it off and put it off, or maybe you dive in and try to learn all of it at once, get overwhelmed, and give up.

The secret to solving this is a combination of being happy to make some messes and a bit of "go slow to go fast." Let's take music as an example. When people learn to play an instrument, they want to play fast like their favorite players. But when they do that before they really understand the instrument, what they mostly do is play a mess: Their timing is inconsistent, and they skip or miss notes. It doesn't sound great, and, worse, they're learning the "wrong way" of playing it so that's what comes out even when they get more experience.

Instead, the right thing to do (if you ask essentially any music teacher) is to **play as slowly as needed to play it perfectly.** That can be really slow at first, and that's okay! Going slow gives your brain a chance to really learn the motion and solidify it, and you'll find that you get to higher speeds faster.

This is related to why kids are "better" at tech than older people. Kids will patiently poke at something, get a small bit of it to work, not be bothered by a bit of mess or mistake, and generally take their time learning the new thing, whether it's tech, language, or music. And this very much applies to programming; It's easy to be intimidated by complex code and try to emulate it, but it's far better to build simple, clean, slow code that you understand and can modify before you dive in and try to be Miles Davis.

Go slow, and don't be afraid of a bit of mess and mistakes — you'll actually get there faster.

The right amount of bullshit

I recently spoke with a programmer friend about some of the new developments in AI such as ChatGPT and other LLMs. We were brainstorming things that you could do with them, and it became really clear to me that he had a much more careful approach than I did — you might say I was "bullshitting" a bit with some of the ideas. Their general direction was good, but some details were missing. We disconnected a bit around that uncertainty.

Neither approach was right or wrong. My friend was right to insist on engineering robustness, something that is absolutely necessary if you're going to build anything. But I think I was also right; it's hard to build anything **meaningful and new** if you don't let your imagination run at least a bit ahead of current capabilities.

So, where is the right balance? There's been a lot of news about companies and founders who got too far out on the "bullshit" curve — Theranos and NXT, for sure. But there are also plenty of examples of things that have at least had a whiff of bullshit to them from time to time that have worked out (like SpaceX).

I deliberately framed this as "bullshit" because I think it's how engineers think about this kind of open-ended thinking. Imagination or exploration might be a better word, but the fundamental idea is the same: If you walk through something one sure step at a time, you'll never be able to take interesting leaps. This is just another way of asking "What if." More specifically, the right amount of bullshit is when you are fairly certain there is a solution to a problem, even if you don't have all of the details just yet. The wrong amount is when you have no idea if the problem is solvable at all or even how you'd approach it, and yet you assert otherwise. And, of course, "too little bullshit" is when you refuse to solve a problem until you have so much detail and certainty that it is essentially already solved.

Fractal boundaries are usually interesting places — where you can't clearly see which side of the boundary you're on, and it changes quickly as you scale and move around. There's one here, between certainty and bullshit, that is likely where all of the interesting, rewarding ideas are.

Salvation is only a half-step away
Improv and innovation

I play bluegrass mandolin a bit. One of the things you're called on to do often is play "breaks" (solos), sometimes for songs you've never heard before. Virtually everyone thinks their break is terrible, and virtually no one actually cares! It's a good lesson in humility.

As hard as playing breaks is, there is some structure to it to help out: You always know what key you're in. That means, out of all of the 12 possible tones you can make in an octave (and there's only "really" one octave; it just moves around), eight of them are the "scale tones" that are in the key you're playing. Those tones will sound okay, if not great, or they at least won't sound too terrible.

The "bad" tones (there are no bad tones) are all fairly close to "good" tones. In fact, that's kind of what makes them "bad": They are too close to another tone, and dissonant. Close in this case means "a half of a step," the smallest amount of musical separation on a fretted or keyboard instrument.

What does all this music theory have to do with software and innovation? Glad you asked! There is a saying — it's in the title above — that when you're jamming, "salvation is only a half-step (meaning one move on the fretboard) away." This is more or less true: Notes that aren't in the scale can sound okay if you keep moving through them to one that is (this is called a passing tone). Sometimes, they give character to the solo: Three of the "bad" tones are blues tones. They make the solo or melody seem sad or moody or build tension toward a resolution. If you keep moving, you can usually make any "mistake" sound interesting or even good.

And this connects directly to innovation and new ideas. We may feel like our new ideas are embarrassing or dumb when we say them, just like a "bad" note in a break. Stalling here is as bad as stalling in a break: It chops up the flow and rhythm. Moving and saying "yes," keeping the rhythm, and seeing what you can do with a partial idea is much more effective and important than being "right," "smart," "visionary," or whatever.

Will you play a sour note or have a truly bad idea from time to time? Yep! But remember: Salvation is always a half-step away.

Don't wait until it's not messy
The right time for new tech

In 2005, when a small team of folks and I started working on Writely, which eventually became Google Docs, the technology around what came to become called "Web 2.0" was in its infancy. JavaScript had been around for a while (in fact, I remember checking how many installed browsers supported it — 80% at the time!), but it wasn't widely used. Just Gmail has made a significant application with it. Browser VMs were buggy and slow; there weren't many systems to support coding (even JQuery came later). On the server side, things were similar: People were building web apps, but not much of what we would call the cloud really existed in a broadly accessible way.

No Prize for Pessimism   63

The answers to these problems were known, more or less, but as an engineer, you had to deal with a lot of mess, flakiness, unsolved problems, and just general hassle. But! It was becoming fairly clear that there was a "there there" to many of us; lots of folks were slogging through this mess and building great solutions and companies.

In the early stages of anything, it's easy to look at emerging technologies, see only the problems, and focus on those issues. These are the doubter's questions that we've discussed here before. They're valid, but they don't really help you as an individual. The better questions are the "What if" ones.

The right time to begin building something interesting with new tech isn't when it's "perfect" or easy. By then, it's too late. The job of an engineer is to solve problems. If you wait until someone else has solved them, you can still add value, but your opportunity to do so will be diminished significantly. The right thing to do is to dive in! And the right time to do it is when things are messy, hard, and uncertain.

There are always problems to be solved with new tech. Some of the problems are real and hard and will shape what's possible, and some are less hard and can be solved with a little care and energy. The key is to not mind the mess; pick something interesting to try to do and wade in.

If everyone waited for it to be easy, it never would be.

In praise of glorious messes
Or "coders code"

I play bluegrass and old-time mandolin as a hobby. One of the things I hear a lot is, "I wish I could play music" (sometimes, I do, too!). I started playing when I turned 50, which is typically kind of late. Many of the folks I play with have been playing their whole lives, but they're welcoming — and I learned to play music, just like them, one note at a time. They welcomed me as a raw beginner.

I also hear people say, "I wish I could code." It's just like learning to play music: You have to get down to it and try. Coders code — it's not any more complicated than that. But just like you don't start playing music by trying for a symphony, you don't start coding by trying to build something large and complex. You learn slowly: one character and one problem at a time.

When we're younger, we have less ego and more willingness to start slow, make mistakes and messes, and be humbled by not knowing things. As we get more capable, we don't like that uncomfortable feeling, and we lose our "beginner's mind" — we become inflexible and stuck.

No Prize for Pessimism   65

We've all heard that the "really great coders (and artists, and musicians, and ... many other things) never stop learning." But there's no magic to that; they just keep allowing themselves to be uncomfortable and to be confronted with new problems. They make glorious messes on the way to being glorious.

Saying "I can't code" or "I could never play music" never helped anyone learn either thing. The only thing that ever does is **trying**: You have to be willing to make that awkward effort that makes a mess and, little by little, you learn. It's a choice: You can never feel awkward and stick to what you know, or you can make a bit of a mess, make a bit of a fool of yourself sometimes, struggle a bit ... and keep exploring the world your whole life.

I know what I like to do.

Lessons from a failed experiment
I wanted high-temp superconductors too!

I've been spending less of my time recently with larger teams, and more with some engineering around AI and LLMs. That's been a good opportunity to exercise some first principles that I write about. It's easy to look at a situation in the past, with perfect insight, and have good advice about it. It's harder to live that advice in the moment.

The world got excited in 2023 about the possibility of room-temperature superconductors. It turns out (it looks like now) that the initial paper was wrong; it's some kind of magnetic effect, and there is no real superconductivity. Bummer!

However, when we were still in the middle of uncertainty, I observed some interesting things. I saw lots of folks falling back on things like references to authority to dismiss the finding. I work with someone who regularly spends time with very serious researchers, all of whom said, "No way, this challenges how we view the world."

They turned out to be right this time, but there plenty of examples of this kind of reaction being wrong. Most really radical paradigm shifts were rejected by the "old guard" for this kind of reason at first. There's even a name for this, Planck's principle, the idea that scientific consensus only changes through generational change.

Why is this? And why should we not just "listen to the experts"? In a word, incentives. The world values new discoveries. But science only gives credit for the ideas that work, not the novel attempts that don't work. No one gets credit for lots of "Well, that was interesting, but it didn't pan out" papers. We barely even give credit for replication studies and negative results! Academic science has even worse incentives; scientists have to play politics and fit in to get tenure, funding, etc. It's really hard for radical ideas to get traction.

But of course, as I've written before, it's much better to have a default setting of "optimist" or at least "open-minded and curious." There is a small prize for being skeptical and right, no prize for being skeptical and wrong, and a huge prize for being open-minded and right! (Okay, you can be open-minded and wrong and waste a lot of time and money, so that square of the graph isn't great, and you do want to have some proof points along the way as you scale up).

At any rate, I don't feel any shame for trying to be open-minded and rejecting the quick "appeal to authority" arguments, even though they turned out to be correct. If you want to break ground and do surprising things, you have to be comfortable with being wrong 20 times before you are right once. But the time you're right is the one that makes it all worth it!

Pragmatic skepticism as a tool

It's fairly easy to be skeptical of something — in fact, like entropy, it's kind of the default state. Entropy happens because there are more disordered states than ordered ones. Skepticism happens because there are more ways something could fail than ways it could succeed (usually). So, it's easy to find a story to be skeptical of if you want to.

But it's not enough to just be skeptical and contrarian. If you're an investor, you have to be contrarian and right. If you're an entrepreneur or engineer, you don't get a lot of credit for just saying "no" — you also have to find a solution you're not skeptical of, and eventually build something useful.

So, it's good to challenge our own skepticism. We can ask questions like "How could this work?" or we can examine the problems we see and ask about potential solutions. When looking at new products, it's often illustrative to look at user value and ask, "What if it works?" rather than "Why won't it?"

And it's good to examine your own biases. I have done this a lot personally with crypto. I'm pretty inherently skeptical about a lot of that field, but I also recognize that my younger self would be happily playing with all the tools and looking for interesting things, so I've been trying to keep myself in that mindset. When I look at solutions in this space that have problems (most of them do), I try to ask, "Can that problem be solved? If it is solved, does this do something useful for a user that hasn't been done before?" Most of the time, I'm still

No Prize for Pessimism   69

skeptical, but some of that exploration has led me to find things that are at least more interesting to ponder further and more than a few related ideas that are probably worth doing.

Skepticism is a very useful tool, both in life and in engineering. But it also has the potential to get in our way and create bias in our thinking if we aren't careful about using it. It should be one tool of many, not the only way we approach challenging ideas.

Mining for ... jewelry?

Everyone wants to be innovative and create the next big, disruptive thing. How do you set yourself up to see these ideas? There's a common misperception that the big ideas are very obvious when you see them. This can be true, but more often, the big ideas are messy, impractical, and hard to distinguish from bad ideas. In some sense, they have to be — if they really were obvious, either you're incredibly lucky (possible, but by definition unlikely), or everyone else will see it too.

As I said in the beginning of this book, looking for an obviously good, big idea is a little like mining for finished pieces of jewelry. Finding a gold nugget in the ground is possible, but even then, you'd likely not want to wear it as a finished piece. Miners are looking for ore, not finished jewelry. Their skills enable them to recognize the difference between an ordinary piece of rock and a valuable one.

You have to have a similar set of skills to mine for good ideas. You have to be able to look past the hard challenges and unfinished bits and see the value. This isn't a new idea; it's the same as the "What if" topic. But it's another useful way to view it: Are you looking for ore or hoping to find a finished piece of jewelry on the ground?

No smooth discontinuities

The title of this one makes it seem obvious. Something can't be both smooth (in the mathematical sense) and discontinuous. And yet, when it comes to software and finding disruptive ideas, we often try to get a discontinuous result with a process that favors smoothness. You can't incrementally change your way to disruption; somewhere in there, there has to be a leap.

However, the relationship between incremental change and disruption is more complicated than that. You can often get close to disruption with some degree of incremental progress. This is sometimes called the "adjacent possible": Lots of inventions come from engineers poking at the edges of what's possible and putting small pieces together. As an engineer, you do have to be very comfortable with making progress where you can all the time.

And once you have crossed the gap to something really new, there will be a lot of incremental work to make it **great**. As an engineer, you can't always be in a "super disruptive" mode. Most of the time, you will be working on more incremental progress.

And yet, those big, uncomfortable jumps have to be in there somewhere. And the discontinuous jumps often seem irrational — after all, that's how you get down off a local maximum, right? You can't increment your way off of it; by definition, you have to either go backward for a while (simulated annealing, which I'll talk about more in the next letter) or make a leap to another domain.

There's no firm rule here other than to embrace mess from time to time. You will often be rewarded for just the opposite, for making small, neat, predictable changes that are to spec. But, every once in a while, a bit of mess and discontinuity can lead to really great things. Think of it as the small bits of chromium and nickel that turn iron into stainless steel: The alloy is more useful than the pure.

Undeveloping conviction

When you come to a new idea, if you're lucky, you'll have a new perspective (at least somewhat new, since you bring your own personal history and biases to anything you do). But in the first moments of dealing with something, you haven't invested much in the problem or solution yet, and you have a lot of freedom to try new things. This is as much true for near-term engineering problems as it is for larger life problems like career, relationship, living arrangements, and job. It's why lots of innovation comes from people who are new to the domain, either new in age (young) or new in expertise (beginner's mind).

No matter the situation though, you quickly settle into some kind of strategy. That's the nature of solving long-term, hard problems or making accommodations in life: You have to make some choices along the way. You begin to accumulate something like sunk costs — the energy and time invested in the current solution only go up. This increases your inertia, which is great if the solution works.

But, sometimes, the solution is a dead end or a local maximum. It might be that you are somewhat happy at your job but not growing. It might be that the solution to the engineering problem works but is buggy and hard to stabilize. At that point, the choices seem stark: Keep going and hope for the best or give up and start over. Most people frame these kinds of problems this way, and usually choose the first path, especially as they get more invested in time or effort.

But there's an intermediate path that's akin to a programming technique called simulated annealing. The key to this is to find a way to relax some of the constraints of the situation. It can be as explicit as actual technical requirements being loosened, but often it's a more subtle relaxation of what is constraining your conviction about the solution. In short: Doubt a little as a way of letting yourself wander around the landscape and get off the hill.

What does this mean in practice? Maybe it means taking up hobbies or traveling to other cities to see if you like them better than the job or city you are in. With an engineering problem, it can mean temporarily trying one of the earlier solutions that were discarded or building something "outside of the design" you're working on to see if it's more effective. Or you can just spend a bit of time poking at the problem from a completely different direction as a "break" from the "real work." Many ultimate solutions are hybrid because of this: The engineering team started with something that was overly simple or constrained but eventually iterated to something that worked.

Just to pick an example, the early history of mobile UX feels like that to me: They started with the (implicit,

74   Sam Schillace

because no one had a better model) constraint of "replicate the desktop," but the best expression of that still didn't really work, so they relaxed a little bit and worked on something that was more like "make it familiar to desktop users." We still have things like distinct applications with icons and there's still a way to text, but scrollbars have been replaced by pinch and zoom, the apps are simpler and more graphically oriented, and you can't quite (and don't need to) organize the device and manage it the way you do a desktop (though you can get closer with Android).

In that case, "desktop but on a phone" was a local maximum that didn't actually solve the problem. "Something desktop-ish but tuned for the device" was an adjacent hill that needed a bit of wandering to get to.

Conviction is good but, like steel, it can be brittle if it's not tempered with a bit of doubt.

How do you go so fast?
"Does talking count as doing something?"

I've sometimes been asked how to execute quickly. In bigger companies, it's typical for things to bog down — there are lots of stakeholders, lots of requirements, products that are older and larger and mature, and all of that. In those contexts, it's understandable that the overhead slows the teams down. The problem is that folks get used to that pace and don't really know what to do when they are working on something new and unburdened.

Another challenge is that folks often confuse **activity** with **progress.** "I'm so tired," "I sat in meetings all day," or "Look at this deck I made!" It's not to say that meetings or documentation aren't useful in defining and building a product, but they're not the same as making concrete progress toward user value or product vision. This is something I see a lot: teams working very hard on tangible output like meetings and documents and research that doesn't actually do much to build a product.

Moving fast can be uncomfortable. It often means doing things before the whole picture is clear, learning as you go, and being nimble about fixing mistakes. It's a different skillset than elaborating large designs in a legacy system. However, the best way to learn and convince someone that your approach is right is usually by using code. Most of the products I've built have felt different than I anticipated and evolved accordingly.

Speed is as much a choice as anything else. It has costs, tradeoffs, and cultural demands. It's not something that just happens; it's a different way of arranging and prioritizing (and rewarding) activity on a team. Every-

76   Sam Schillace

one wants to be **faster** — the question is whether you also want to give up some of the things that are incompatible with that and do what it demands.

Learning by doing

I spent a few days last week (October 2023) on the campus of the University of Michigan, my alma mater — always a fun and humbling thing to do. The students there had a lot of good questions, and we had some good conversations.

I get asked a lot about how to build a successful career, and I never really quite like the answers I can give because so much of my own career was lucky or privileged. I will often give advice in the direction of overthinking, which I believe a lot. So, for example, if you want to be a coder, well, coders code, so write a lot. If you want to build products, you have a lot to learn from the market, so make it easy to do that: Get your tools sharp and make it easy to conduct experiments quickly, things like that.

So I was happy when I was in one of these conversations with someone, and she pointed out some actual research in this direction. It was an experiment where two groups were asked to make pots. One group was told to make as many as possible, and one group was told to make a single pot but to do it as well as possible. Guess which team made the best pot? Yep, the one that practiced a bunch and made a lot of them.

There's a lot to unpack here. Why did the first team do so well? A critical part of it was that they were **given permission to experiment and fail.** The instructions contained nothing at all about quality, just about making pots. This meant they were just focused on the process and learning by doing and didn't have any internal editors second-guessing them.

You can do this for yourself, too! I often tell people, "Just follow your nose where it leads." Again, I was lucky that this worked as well as it did, for sure, but it's a really good technique for letting go of that inner editor and just building things. In fact, one of my favorite parts of any new, disruptive area is that **no one** is an expert yet, so it's by definition okay to just try things and see if they work.

But of course, any area is always new to you when you first encounter it. If you can get into "beginner's mind" and let yourself just work without judgment, you will often find that you get better results more quickly.

We are in a moment like this with generative AI. We might not know whether or how it is intelligent or useful, but there is no doubt that it's complicated enough for there to be a **debate.** No one has the roadmap. You will likely get further by trying to build lots of things and doing lots of "weird" experiments than you will by sitting in a corner thinking things out until the time is up.

Make pots, lots of them. That's the best way to make one perfect one.

Chapter 4

Technical Philosophy

A toddler on an incline
And the rainbow of disgust

One common coding experience is dealing with a codebase that could be better designed for the task at hand. It's very easy to second-guess the initial design — "We should have taken more time" or "That was never a good design," etc.

But it's usually with the benefit of hindsight, sitting on a successful product, that we have that perspective. It's closer to reality to understand that the code had to get written fairly quickly to get to market ahead of competitors. In the moment, we didn't know everything we know now. It's very much as easy to fail by being too slow and "perfect" as it is to fail by being so fast and sloppy that the code melts down.

Failing to navigate this tension between speed and maintainability is one of the primary ways to fail in a software project. Many successful products and companies go through a period of crisis when the early hacks fail to scale as the team or traffic gets too large. Spotting this and doing refactoring and maintenance early enough is a critical part of every great effort.

So, how do you understand whether you are being too slow and neat or too fast and careful? Enter the "rainbow of disgust"! There is a spectrum of emotions you can have about your code, roughly in some order from bad to good: terror, disgust, discomfort, embarrassment, calm, happiness, and pride.

The right place to be on that scale is roughly the middle: slightly embarrassed but not afraid, disgusted, or terrified. Another way to visualize this is, if you've ever had to be in charge of a toddler, is the sense of them running down a slight incline. They're not entirely in control, but also not entirely hazardous — something to be watched and managed but not stopped.

Most projects that are successful have this sense of pragmatic tension between the need to get things done quickly and the desire to make the code as clean as possible. There are lots of reasons why "too clean" is a mistake, mostly because you won't understand the problem as well as you think until you've solved some of it. And, of course, the right place to be on the rainbow depends on the project; for some kinds of software, "very slow and correct" is the right answer.

But more often than not, "slightly embarrassed and moving fast" is the right place to hang out.

Don't lie to the computer

This seems like a funny thing to say. Why would anyone ever lie to a computer? To spare its feelings? Social awkwardness? It's absurd!

But it actually happens more often than you think, and it usually ends in tears. Computers, as we all know, are really dumb. They'll take you at your word and thus won't be able to do the right thing.

This happens sometimes with things like bug databases and alerts. I've seen teams do things like invent "bug level 2.5" because the level 3s were too messy to deal with, but they needed to fix some things that were not really as bad as 2s. This is lying to the computer! You're pretending there's really this other level and that it's okay that there are so many 3s, but it's not true.

This can happen in software design, too. Usually, it takes the form of not really thinking clearly about the fundamentals of the problem at hand. It could be that you've designed something that is $O(N^2)$ and are ignoring the fact that it's possible to have really large inputs. Or it might be that you're pretending that some algorithm can compute something in all cases when it's really only working in the current case you care about. That would be okay, but if it gets labeled "general thing," that's lying, and, at some point, the computer or a fellow programmer will fall over because of it.

It's always tempting to lie to the computer. We often don't have time to think through all of the edge cases and implications of a design and we're impatient, so it is really tempting to be dishonest about what the problem really is so we can ignore parts of it. However, as with any other relationship, lying is usually not a good idea in the long run.

Noisy monitors

One of the most pernicious and common patterns that engineering teams fall prey to is the "noisy monitor." This is any kind of signal — a pager, compiler warning, or bug database — that accumulates noise relative to useful signal over time. This pattern is often the cause of outages and product failures. The problem with noise is that we are pretty good as humans at weeding it out, and we are very good at making up reasons the warnings we are getting "aren't important." Sometimes, we build tools for this, like silencing a pager or scripts that filter out the "noisy" errors, and sometimes we just acclimate to it.

There are lots of ways that noisy monitors cause problems. I worked with a team once whose pager went off about 2,000 times in a "normal" day (ugh, right?). No problem, they just "stfu 10" every once in a while, to quiet it down — except when they had an actual outage and had to dig though about 1,000 new pages **a minute** to see what the issue was. There was so much noise that the signal was entirely lost.

But there are more subtle ways that noise in our monitors causes problems. Bug databases are another example of this phenomenon. Most teams gradually accrete a large pile of old, low-ish priority bugs over time. Or even if they're not low priority, there's a temptation to say, "Well, we haven't fixed that in two years; it must not matter." But this obscures both newer quality issues and the overall drift of the product. It's far better to resolve something into the right bucket or "will not fix" to keep the tool clean for the things that matter.

Compiler warnings are another great example where this shows up. The warning is there for a reason, but we usually decide it doesn't matter enough to fix the code to not throw it. So, we silence the warning and then it doesn't fire when we hit a case that matters. This is a case where zero tolerance is the right approach.

I read a story once about how Thomas Keller, chef of the French Laundry, would come in first thing in the morning and clean and organize all the bottles and small tools. Every day, no matter what, he would reset the kitchen to the base state as cleanly and consistently as possible. I've never seen a software team be that disciplined (at any company), but we ought to be. To keep things that well-organized requires a high degree of clarity about the problem being solved (and what's important about it) as well as discipline not to be distracted — or let monitors get dirty.

To hack or not to hack
The benefits of general vs. specific

One of the most durable and difficult problems in software development is choosing carefully between building something in a general, modular way versus building for just the specific problem at hand. This can sometimes be described as "future-proofing" or sometimes just "hacky" or not.

The challenge is that neither pure approach works completely. If you just hack everything, eventually, the code gets big and messy enough that it's not manageable; it's a "hairball" in technical terms or the dreaded "mudball." If you imagine a graph of coding speed over time, you did great at first but then dropped off a bunch, so the area under the curve isn't that big.

But! If you are overly focused on modularity and understanding every possible future use, you often go much more slowly than you think. It's possible to get good leverage from *some* generalizations, but often, the effort of building the generalization doesn't pay back in the lifetime of the code, or you build the general case before you understand the problem well and some of the effort is just wasted. If it takes 3 times as much energy to do something generally, but you only need two instances of it, you didn't come out ahead. It's even worse if you need zero instances of it! In this case, the area under the graph is small at the start but, hopefully, gets bigger over time.

And that's really the right way to think about the problem: What's most likely to maximize the area under the "code produced over time" graph. Sometimes it's worth delaying at the start to have a payoff at the end, sometimes it's not. The two questions to ask are: How hard is this to change later (if it's too hard, then you're likely too hacky) and how long until the modularity pays off in reduced total effort (if it's too long, or if you don't understand the problem well enough to even have a guess at ROI, you're probably being too careful).

No Prize for Pessimism   87

It's a tough balance, and there are legitimate times when the answer is unknowable and is just a judgment call. One specific thing you can do as an engineer is keep track of your predictions as honestly as you can, try to pay attention to how you make these calls, and look for where you can be better at them. It's a lifelong process.

Language isn't the cure
Systems and incentives are closer

One of the common patterns in programming is the idea that a new language or programming technique, like functional programming, will magically cure all the things that annoy us. The most dangerous time to fall into this trap is around the 10-year mark; we are seasoned enough to have a long list of annoyances, and some experience with broken projects and tech debt. We're experienced; let's show the world how it's done! This approach usually doesn't work out well.

That's not to say new languages, libraries, and techniques can't be better than what they were before or that we should never adopt them. Obviously, we have to improve and innovate as an industry, so someone has to adopt new things all the time.

The problem is when teams and engineers expect a new language to fix old systemic problems. Don't like writing tests or documentation? No matter what it claims, there's no language that will really fix that problem for you except at the lowest level, like memory leaks. If your code is poorly organized or designed, the language can't help make your thoughts better; only you can. It's like saying, "Well, I'm a terrible writer in English, but Spanish sounds so romantic; I'll learn that and then be a great novelist." It's not very likely **and it's** a lot of work.

Systemic issues on the team are similar. No language can prevent bad patterns such as hero culture or lack of trust or can help you make good choices about level of tooling investment, good code ownership practices, tech debt cleanup and maintenance, noisy signals, and the rest. That's hard work that you have to do before you write any code.

What's worse is that if you make the wrong language choice, you can not only be stuck with the same problems but also with code written in a language that no one really knows or that is hard to hire for. It's easy to come out behind by doing this.

New things are fun to learn and are sometimes better approaches. Usually, though, the mundane work of discipline, craft, care, and systemic incentives with a boring mainstream "workhorse" will get you where you need to go faster.

Solve the right problem
Understand, don't over-optimize

Engineers like to talk about the concept of "tech debt." Sometimes the metaphor really makes sense: Neglecting maintenance on code sometimes accrues like debt does, making the engineering team slower. But there are other kinds of tech debt that are more subtle and harder to address. One of these is unnecessary complexity. Over-designed code can be as hard to maintain as under-designed code.

This can come from many places; it usually has to do with not clearly understanding the problem and solving a more general, but harder, case. Sometimes it comes from not making hard choices about user experience and trying instead to make those choices in the code.

A couple of examples:

In the early days of Writely, we had a fairly complex algorithm for doing merges. It was hard to maintain already, but we started to get into some edge cases that made it really hard — what if two or three users were editing on the same line at the same time, and one of them deleted an edit from the other? The state space here is complex — we never want to lose data — so what do we do? After bogging down on this problem for a while, we finally decided the answer was: "Don't do that."

90   Sam Schillace

The problem here isn't the software; it's the humans. You can't expect mere software to stop humans from fighting, so we didn't optimize for that case, and it got easier. We added presence instead so people could work it out.

Another favorite example is from the early days of Google Spreadsheets. They had a fairly challenging latency problem because the numerical models are complex enough to get on and off disk, and it's hard to fit that into the amount of time you have for a web response. So, keeping it in memory makes some sense, but it's hard to update a complex in-memory model in two datacenters within that latency budget, too. The solution they came up with was elegant: Since failover is relatively rare, tune the system so that the replication is fast (just some log entries to disk that can be recombined later), but if you need to bring up a replica in the other datacenter, that's a bit slower. It's not quite as optimal, but the code is simpler, and it covers the common case well.

Being thoughtful about the product requirements that make your code more complex and removing as many as you can is an important aspect of minimizing tech debt. As engineers, we often want to solve all of the details of the problem (which is pretty much our job), but the real goal is to have the software work for as many people as well as possible. Sometimes, narrowing the design and making it simpler is the right way to achieve this in the long run.

What guarantees are you making?

Reading postmortems is one of the fun things (really) about being around large services. It's always interesting to play amateur detective and dig through what happened to take a service down. It's a little like true crime but with software.

One of the patterns that come up a lot in these failures is that there was an aspect of the system that was being depended on as though it was a guarantee, but it wasn't a guarantee — and it changed. For example, a specific routing behavior might be said by the documentation to be random, but it might always be random in the same way (so not random) until something changes in how the order is added (it's random now!). Sometimes this happens when a new replica or route is added to a system, for example, or there is a partition or even sometimes a deployment.

There can also be side effects that are silently dependent; a read might not usually cause a commit or a cache flush but then will suddenly start doing it and causing load or latency issues. Sometimes the guarantees are about latency: Code will assume that a remote procedural call

(RPC) will return quickly even in the error case, which it does until something on the other end is slow rather than down, or the cache is messed up and missing all the time, and it takes 30 seconds instead of 3 milliseconds.

When you build something that's going to be used by someone else, stop and look at it from outside. What are you guaranteeing? What would a reasonable person *think* you are guaranteeing? Are you? Can you make it explicit? Can you describe "anti-guarantees," or things you aren't guaranteeing? Being able to see and describe these is an important part of building reliable systems.

Fast vs. right
What can you fix later?

There is a well-known (and old) paper called "The Rise of Worse is Better." That paper describes some of the surprising ways in which technical adoption happens. It discusses the importance of simplicity versus completeness and correctness, making the case that something that is simple but not complete (and sometimes not correct) will often beat something that is "better" but more complex.

To a large degree, that pattern is about speed. It takes time to get something completely right, and it takes time for programmers to learn something more complex. Something that is simple and gets the job done today — and maybe has problems to solve tomorrow — is often a winning pattern.

The tech industry is cyclic. We are in a part of the cycle (in March 2023) when many new patterns and standards are being discovered and deployed. This is a "disruptive" phase rather than the "incremental" phase we are more used to. In this phase, speed is even more important than ever. It is often the case that the first person out the door with a new programming paradigm, app model, or business pattern is the long-term winner just because they got started first. Sometimes, the difference is measured in weeks.

This is particularly true with things that have network effects. Networks are inherently exponential: Once a network effect starts, even at a small scale, it almost instantly becomes nearly impossible for someone else to disrupt unless they can immediately leapfrog with some other injection of use.

One good way to think about this is whether something can be fixed later or not. Some things (like minor syntax details) can be fixed later, but others (like ecosystem momentum) can't. It's a fatal error at this moment to worry about things that can be fixed later if that causes problems that can't.

This shows up differently for different teams and companies. It's a hard mind shift to make, particularly after a long period of mostly incremental improvements that have valued caution and correctness. The tricky thing about this kind of moment is that you also don't have much time to recognize and react to it.

Dangerous productivity metrics
What's the worst that can happen?

We all know that it's notoriously hard to measure developer productivity. Almost any concrete metric — lines of code, bugs fixed, pull request counts, etc. — will, at best, tell an incomplete story and, at worst (the usual case), be gamed.

In fact, giving an organization a metric like this is a bit like overfitting an ML model. It's important to be very careful what you are "training" on, or you'll get strange side effects.

Usually, when managers look at this kind of measurement, they ask what seems to be a reasonable question: "What's the best metric for measuring some aspect of performance that I can think of and implement quickly?" This is actually a very dangerous question and likely the wrong one.

The first question you should ask about any metric

you want to apply in public to your teams is: What's the worst possible misuse of this that I can think of? Because, sadly, it's highly likely that something approximating that misuse will be at least a side effect. Count pull requests (PRs)? Expect to get a lot of them and be utterly unable to distinguish signal from noise forever. Bugs fixed? Expect your bug database to explode with trivial bugs disguised to look like real issues, completely removing your ability to get a good read on quality. And so on.

This issue is similar to overfitting, and similarly defeats its own purpose. Organizations do learn, and they do respond to inputs. For example, it's valuable to have check metrics, which function as test data. But it's important to make sure the inputs actually reflect what you want to train on (input "count PRs, train on 'PR count is important'").

In AI, individual nodes aren't smart, but a bunch of them together can be. In organizations, it's almost the opposite: Individuals are smart, but collective emergent behavior can be quite "dumb" in that it will sometimes take instructions very literally.

Developer productivity will never be fully resolved. But as you approach it in your teams, be careful to consider the side effects of the metrics you choose.

Chapter 5

AI and Products

Intent and iteration are the new click and drag
AI-based UX

The current generation (April 2023) of LLMs is beginning to let us move from syntax and process to semantics and intent. This is incredibly powerful and valuable for a bunch of reasons. The intent is much higher leverage and much more robust than process. Saying, "Go into the kitchen and get the milk," is denser than saying, "Lift your left foot 3cm at .5cm/s, swing forward at 4 degrees per ms," etc. And a precise process for walking into the kitchen to get the milk gets even bigger and harder once you have to write all the edge cases ("if cat present, then"). It's not feasible.

The intent has a higher density, but the underlying system has to be capable of understanding and executing that intent. That's where there are some opportunities in recent LLMs such as GPT-4. But there are still challenges here. One of them is precision: The models are very rich and very capable but not always repeatable and not always accurate (famously, they hallucinate; more on that later).

At the intersection of these two ideas is an approach to designing programs and UX that sounds simple, but it is probably as foundational as click and drag has become: iteration. We aren't used to designing this way. Oh, sure, we iterate a lot on our own content; we rewrite posts and images, try things, and back out and go again.

But we're not used to doing that with something we think of as a program. You don't get to change the icon you clicked on if it doesn't do the right thing. It's static, and it either works or doesn't. But in the realm of language or meaning, things aren't that simple. Maybe you didn't express yourself well. Maybe the compression worked too well, and you need to clarify ("Which milk? Oat or almond?"). Maybe you were mistaken about what you wanted and needed to correct ("Actually, maybe I'll have tea.").

There will almost certainly be best practices that emerge, just as there have been for click-and-drag interfaces. There will be a balance between giving the user breadth and giving guidance, just like there is a balance between having lots of functionality and having too cluttered a graphical design. And there will probably be both good and less good practitioners of design in this space of iteration.

I believe that most of what we will eventually do will be "talking" to the computer — where talking right now is mostly defined as "text chat" but will shortly be multimodal: images, gestures, voice, etc. As those interactions get more complex, it will be harder and harder to build "rigid" interfaces like we do now. Why build a bunch of static buttons that reflect an underlying, rigid schema somewhere when you can let the user tell you what you want, find the result in a vector database, and iterate together?

Intent and iteration will be a foundational metaphor for user experience in the next wave of software. We got click-and-drag Windows interfaces when the tech was advanced enough to give us high-resolution screens and fast processors that could handle real-time interaction. We now have new capabilities that let us handle interaction with intent and meaning in real time. It's time to build the experiences that are native to those capabilities.

Pixels are free now

Well, not quite now (September 2023), and not entirely free, but soon and almost.

What does this nonsense mean? Before the internet happened, many businesses were predicated on the idea that certain kinds of distribution had high costs and friction. But when the internet appeared, distribution became "free," and any business that depended on that — managed it (like retailers), amortized a bundled package of businesses with it (like newspapers), or took advantage of it to be gatekeepers (like the media industry) — was disrupted and reinvented. Many resisted, failed to adapt, and are gone, but basic economics fundamentally changed, and the world changed with it.

Pixels still have high friction, just like distribution used to, but this is changing rapidly, and the end state will likely look a lot like the change that happened to distribution: They will be mostly "free."

What does it mean that pixels have "friction"? It means that anything that puts a pixel in front of you has fairly high costs now and lots of layers of a stack behind it. It's easiest to see with digital images. A year or two ago, a pixel in a digital image had thousands of hours of Photoshop engineering that went into the tool that the user then spent their own hours on, learning and then drawing. But now, almost anyone can get a generative model to generate those pixels (or "good enough" versions) much more cheaply. Those pixels are becoming "free" in the sense of low friction (and yes, I'm leaving out artist compensation here — it's a real and important issue that is separate from what we are discussing here. Even for artists, though, pixels will become "free" as the tools get better with AI. And, just like other digital media, the market will get bigger and flatter. It's super easy to produce music or video now, much harder to get attention, etc.).

This isn't just true for digital art, though. Imagine an HR lead who gets asked to do a salary analysis. This used to be high friction and slow: Get an engineer or a data analyst, build connections to the data or sign up with a SaaS company that has built a bunch of expensive code to do that, and spend weeks on the task. But with AI and things like code integrated into agents, even simple ones like ChatGPT, this task is much faster and easier, taking days or sometimes hours instead of weeks. The pixels of the final report just got much cheaper.

It's not hard to imagine that "applications" or documents in a few years will be more like conversations: Tell the assistant what you want, and it does the drawing and builds the UI for you. In fact, in the early days of GDocs, I predicted (easy prediction) that anything not connected to the internet would soon seem "broken" (this was pre-smartphone, so it was surprisingly controversial). This is going to happen for something like "smart": If your agent can't talk to an app or a business, if you can't ask a document to "show me a graph explaining this," or if someone expects you to push pixels around a spreadsheet — it's all going to seem "broken." Very soon. Pixels will become free, and we will become irritated and then annoyed and then route around the places where they aren't.

(An interesting side question here: What are the primitives that this rests on stably? And how will applications "unbundle"? Albums broke away first from physical media and then moved into streaming, where we are more focused now on tracks as the primitive. Newspapers broke into multiple different businesses and markets, and many more small niches are now addressed by things like blogging [and other] platforms. What happens as applications unbundle from fixed UI, as operating systems become more like conversations, as databases can more easily handle ad hoc queries and combinations? What's the next settling point for each layer of the stack?)

There are many challenges here and much to build. The point of this letter is not to cheerlead this change but to call it out so we can think about it. It's not likely that we will, collectively, decide we want the friction to stay; that almost never happens in any economic activity, at least for long (sometimes gatekeepers hold it at bay for a while, in some domains, but people eventually figure out there is a better way and pressure builds). At best, there will be niche markets like the current market in vinyl, but the mass market, the bulk of users, will always choose the easier and more convenient (which usually means the cheapest) option.

Pixels will become cheaper over time. What are you building that could be built in an easier way, one that's better and more flexible for the user? What "toy use" or small niche market you serve has already started to figure this out, and built your disruptor?

AI and the end of integration testing

Every fundamental change in how software can be delivered or built results in a corresponding change in the toolchain and common practices for building it. The desktop had waterfall and manual testing, the cloud had CI/CD (continuous integration and continuous deployment) and automated testing, and there was more dependency on telemetry. AI and LLMs will be no different.

We are used to being able to perform both unit and integration tests on software before we release it. Both involve defining some fixed environment and domains that we can test inside to look for the behaviors we want. Integration testing, particularly with things like large-scale distributed services, is notoriously hard because you have to simulate a large environment and many interactions very accurately to get a meaningful and complete result. Programmers still do a lot of "smoke testing" and manual "checking" because it's hard to admit that integration testing anything fully is hard. As an industry, we rely instead on telemetry and user reporting to catch what we miss.

AI, particularly more independent agents, is going to finish breaking our capability to do integration testing in advance and is going to usher in a different kind of development pattern, or at least evolutions of the current one.

Why is integration testing going to become impossible? Largely because the aperture possible for a program (or digital agent) is essentially going to get infinitely wide. We are already seeing some early attempts at this. If an agent is "autonomous" and roughly as capable as a human, then the list of other agents, programs, services, and actions they can interact with is essentially as infinite as the natural language that drives it all — as complex as the real world, and impossible to fully simulate. Even if you could set up a full replication of the real world as a sandbox to test in, the combinatorial complexity is far too high to test even a representative sample.

What does this mean for testing? How do we build and operate safe software in a world where we can't test as much as we want? I don't know — no one fully does — but I think this will drive us to be more telemetry-focused and to build "self-checking" systems that will cost more in compute but do more checking and correction at runtime instead of at test time.

This is hard to do. We aren't talking about measuring things like latency or crashes but about measuring more semantic properties like "safe," "helpful," "nice," and "making progress." These will likely require their own classifiers and inference to do well, which, of course, is a hard security problem at scale. We will have to learn to use a lot more computing to monitor and manage agents at scale; single prompts can be mostly tested in isolation now, but agents that are more complex won't be testable in the same way.

We call this idea "semantic telemetry" because of the need to test semantic properties in real time. It's a challenge! There's no absolute measure of, say, "helpful." There can only be examples and rubrics and, hopefully, a stable relative measure on some fixed scale. It might be the case that, as an industry, we will produce common behavior rubrics and start to do things like certify an agent holds to them — it's hard to say.

There are other testing challenges coming, too; regression testing is another one that seems apparent and similarly complex in the semantic realm. Because so much of what these might do is open-ended, it is likely that anything that is highly dependent on fixed behaviors of the base model will be too brittle, but there will still be things like verbosity or maybe "basic intelligence" that more complex programs will depend on. How do we specify and test this? Is there a real-time component where we can make predictions about which model will work right and schedule an inference accordingly ("semantic scheduling" and optimization)?

There are likely even more problems that will emerge as larger teams begin trying to build more and more complex programs and agents that use LLM base models and other AI models as programming objects. The era of semantic engineering is beginning, and we will have to find new development patterns that work for it.

Building applications with LLMs
The Semantic Kernel (SK)

When a small team of us at Microsoft got access to what the world knows now as GPT-4, we learned that it has pretty amazing capabilities! We immediately set about building whatever we could with it.

But we immediately ran into a challenge: As great as it is, at some level, it "just" takes an array (of text, or increasingly, binary data) and "rearranges" it. That's a great function call to have, but it's "just" a stochastic, pure function; there are no side effects, no state, no callouts. It's hard to build a complex program with just one function!

So we decided to start looking at programming tools that would give us some of those capabilities: memory (state), procedural control where we want it, and seamless interaction with native code so we could do RPCs and have other kinds of side effects. This was the beginning of the Semantic Kernel, software we put into Open Source on March 1, 2023. You can find it on GitHub at microsoft/semantic-kernel.

The SK has some basic behaviors that you would expect: We organize things into Skills and Commands that are very "unix-like." They can be written in a programming language like C#, Python, or Typescript or in a prompt-template format. You can pass input and parameters in and out of them and chain them as you'd expect. Individual templates can be configured to use different models and settings, which comes in handy when building larger and more complex behaviors.

We also built a memory store that uses the same vector embeddings that Open AI publishes. This lets us store and retrieve semantically rich memories. The whole point of this kind of programming is to get into the "messy" realm of semantics — meaning and intent — so we needed a memory store that worked that way.

That quickly led us to an interesting idea: GPT-4 itself is so capable that it can actually be a partner in this system. It can be a programming component itself. If we described all of the skills in the system with natural language and stored them in the same kind of vector database, then we could also write "planner" skills that can be used to examine the capabilities of the system and build their own programs. One of the first ones we did is called "contextQuery," which simply categorizes questions in terms of the command or action needed to resolve them; this can reduce the tendency to get "hallucinations" in some cases. There are others we've built and will continue to build.

We're starting to think about what a common language or library of skills would look like. Many problems seem to break down into "hierarchical planning," where there are successive levels of planning and resolution. Can we make this a general enough pattern that the models can make use of it? There are common patterns with how prompt templates get written that seem to make them behave better: ways to break a task down, performance optimizations, and things like that. Some of those are already in the repo, but we expect more to emerge as we and the developer community learn together (as a side note – this era reminds me of the early "Web 2.0" era where we were all trying to figure out the right patterns for building apps in the browser, building distributed services, building mobile apps when the iPhone appeared. I think we are in the middle of the same kind of industry-wide conversation now about AI programming. Not just building and training models but how we work with them as application and service builders. It's fascinating now, just like it was then).

We are also beginning to build out what we think of as "connector skills" for tools and services in the Microsoft ecosystem. We want to be able to build richer and more complex experiences out of the familiar tools we use, and we know other developers do as well. Over time, we will continue to add to this collection.

We've been having a lot of fun building larger, longer-running projects with the SK. I wrote something a few weeks back that takes a short prompt like "eighth-grade history in a friendly style, taught to someone who likes analogies" and turns it into a full-length textbook and curriculum, including a teacher guide, table of contents, the whole thing. There's an art to combining procedural code (loops and such) with LLM prompts. Each is good at something the other is bad at, and so crossing the boundary successfully is a new discipline. But it's worth it! We can build really interesting things with just one prompt call — imagine what you can build with 1,000.

Looking forward to building and learning together. Have a look, give some feedback, and make some cool stuff!

We have always wanted to talk to the machines
Now we kind of can!

I've been thinking a lot lately (March 2023) about the user energy and excitement that is going into AI-based chat interfaces. I think we all understand pretty quickly why these experiences feel so compelling, but at some level, it's curious; we live in a world with lots of rich media: tons of video, games, images, etc. Why are we so drawn back to the "retro" experience of plain (not even formatted) text?

I think it's because, since the actual dawn of the computer industry, we've been trying to talk to computers directly. It helps to think of them almost as this Star Trek style "alien species" that can only communicate in binary arithmetic. There was something compelling and rich there, but it saw the world in a very different way than we did, and we needed to figure out how to communicate across that gap.

So we've spent a hundred years trying to develop a common language so we could speak directly to this new "species." We started with binary codes and math, logic gates, and things like that. Then we moved to assembly, then programming languages that looked a bit like our own languages but were structured enough that the machines could understand them, with some help from translators that we called compilers and, later, interpreters (hmmm). The languages got more abstract and more sophisticated. Layers kept getting built that let us express more complex ideas more naturally (in theory, at least).

And now we are getting really close! We can't quite get the machines to do everything we want just by talking to them, but we can get them to do a lot. We can speak fairly naturally, and they sometimes seem to understand. There are still errors in understanding and miscommunications (hallucinations). But these chat experiences feel much closer to "just talking to" the computer. A non-programmer can express a complex intent to a computer with a decent chance of getting what they want without having to "write code."

We still have a long way to go, and many problems left to solve. But it seems now that this is what we have always been pushing for, in some way: the ability to "just tell the computer what to do, and it will do it." That's why "plain text" is exciting again.

AI is a funny co-worker

You could describe the current crop of AI tools in a somewhat funny way. They're like having a co-worker who:

- is super smart and well read, about absolutely everything, is very fast to respond, but is a little bit of a bloviator and sometimes makes things up;
- is always around on chat and email, and is willing to do absolutely anything (within safety reason) you ask at any time, without complaining;
- has a great attitude and never gets bored or frustrated or angry;
- but really doesn't do anything other than chat or email, even though they have lots of opinions;
- doesn't have a great memory from day to day and often has to be reminded of basic context;
- and if you criticize them, they don't get frustrated, but they redo the entire task from scratch each time, so it's really hard to get them to make fine tweaks.

(Although they are really happy and enthusiastic about making those changes.)

Imagine that in a person, for a second. It's weird! Even though the person is super smart and wants to be helpful, they'd probably be really frustrating to work with. You'd think "Man, when they're on it, it's fantastic, but so often I have to do work to keep them from doing their weird stuff, and why can't they just go do the task instead of always giving me great instructions?" You probably wouldn't hire someone, no matter how smart, with this description.

But if you had them, you'd probably coach them, right? They're so close! Let's teach them to write things down so they remember, let's find a way to get them to make small changes instead of starting over, let's get them trained on and signed up for some tools so they can do some work themselves. You can see the path to them being a great co-worker. And once they are that great co-worker, you can make copies and have as many as you want. So, there's a lot of motivation.

Current LLM tech has limitations, and it's hard to combine stochastic outputs with deterministic code, for sure. From the example above, though, you can see clearly that if we work out the plumbing, these have the potential to be really fantastic. The challenge is that plumbing has to be really easy to consume; you wouldn't want a coworker who constantly needed you to write code to interact with them, or kept forgetting how to send an email, or what its job was, or whatever.

From the perspective of the average user (probably not you, if you're reading this), the example above is illustrative — it has to feel like that! It really is like a co-worker with, perhaps, some quirks, but which can be interacted with casually, naturally, and which can get real work done over long periods. This is hard, and most of my work for the past 18 months has been focused on it. We collectively are making progress, and we have lots of work to do as an industry. But I believe strongly that there is a huge amount of value waiting to be created just from really basic "blocking and tackling" of product and interaction design. And that's before whatever new model capabilities are on the horizon!

AI isn't a feature of your product...
Your product is a feature of AI

(The idea for this came from a February 2023 conversation with Manik Gupta, who said I could steal it!)

As people wake up to AI's potential across the industry, we see a common pattern: People "adding some AI" to their products or processes. But, as the title of this post says, that might not necessarily be the right way to think about it.

In general, as new paradigms emerge, we think in terms of familiar patterns. In the early stages of the internet, companies that had been building desktop software took desktop-centric approaches to the internet and built things like web page editors, browser plugins, etc. However, the real energy and value came from treating the internet as a platform. Amazon, eBay, Google, and Facebook don't have "some internet as a feature of their product." Their products are fundamentally built for the internet. They don't have the internet "mixed in" — there is no product without it!

I believe that AI is, or will become, another platform at the scale of the internet. It's in its early stages now, and there are plenty of challenges and technical issues to be dealt with, but setting that aside for a moment, the same pattern is very likely to hold here. It will be possible to add some value by building AI into your product, but the massive value will come from building apps and solutions that won't work at all without it.

We don't know the shape of those apps yet, and we are barely at the point where we understand what the platform is, how to use it, and what the best coding practices even are. All of this was true at the start of the PC, internet, and smartphone platforms, too.

As always, this is a good time to ask the "What if" questions. What if this really is a new platform? Assuming it all works well, what would you build?

Syntax vs. semantics

There are lots of debates about whether the current LLM approach will lead to "true AGI" (though I suspect we are going to see a lot of "no true Scotsman" thinking about what intelligence is, as models get better). But I do think there is an important shift happening in the nature of coding itself because of these models: the shift from syntax to semantics.

One of the arguments about LLMs is that they are just "stochastic parrots" that have no real understanding of the world because all they do is understand word probabilities. I think this is a shallow argument. Natural language embeds a great deal of understanding about the world, almost by definition. If I say, "The monkey ate the," you are likely to complete that with "banana" or "grape," not because you have done some math on a fundamental property of the letters in the sentence (aha! the binary is one mod 13 so the next word must be 0 mod 11! or whatever), but because there is some knowledge

116   Sam Schillace

of the world and the culture you live in embedded in our usage of language.

We can begin to see this in some of the more advanced models. There are many examples of errors and misunderstandings, but it's also the case that we can see places where ChatGPT and others have some grasp of meaning — in other words, semantics.

For the entirety of the computer industry, we have been forced to deal with process and syntax. We have to mediate between the complexity of the real world and the rigidity of the digital; we build schemas and algorithms. But in our daily lives, we don't live that way; we are iterative, goal-driven, and reactive. We deal with semantics instead of syntax and goals; we deal with intent instead of rigid processes; and we do iteration and react instead of having fixed algorithms.

We aren't all the way there yet, by any means, but this is a fundamental shift in how software will be conceived of and built for the rest of our lives. This is the "second version" of the computer industry. Of course, there have been folks working in this direction with some results for years, but now we are at the point where this fundamental shift is becoming widely accessible.

This is very analogous to the early internet, which existed for many years in smaller form until enough

No Prize for Pessimism   117

new technologies (better packet routing, faster computers, HTML, and the browser) came together to make it widely accessible. There was a quick period of excitement, some early missteps, and then a long period of really figuring out how to make everything work.

In the early days of the internet, there was a debate about whether things like e-commerce would work, whether it was worth it for a business to have a website, etc. Now, it's fairly obvious that if you aren't accessible to the internet, at some level, you don't exist. The same is going to happen for software and companies in the age of semantics and AI: Anything that isn't available to AI, that can't be managed easily through non-brittle goals and iteration, will start to seem "invisible."

Coordination efforts
Humans, work, play, and AI

Usually, these letters result from something I've observed during the week at work. I took a few days off this week (May 2023), so less was happening, but I saw something interesting nonetheless.

We are all familiar with the idea of the mythical man-month: the idea that software teams get less efficient as you add people because of communication overhead. This is generally held to be because software is so intensely communication-centered, but I think that's not the complete story. It's also interesting that we've had something like 40-plus years of innovation in programming tools — libraries, languages, frameworks, app engines, different database techniques, better debugging, websites like stack overflow, now LLMs, and on and on — and we still seem to have this problem.

I was at a vacation house with my family. I do some work around this house when I'm there alone, and I'm very efficient. I often wake up in the morning with five or 10 things to get done, and most of them are done by the end of the day. My sister and daughter were here for the weekend with my sister's family, so four to six people, depending on the day. We were much slower to organize, and when we had that many people in the house, we were lucky to do two things in a day.

I don't know why this is. I think it's at least partially social; no one wants to be the "boss," so (to put it in server terms), instead of having a "primary," we have to fall to some kind of consensus mechanism, which is more expensive. I think this is just durable with any group of humans making a choice: It's either hierarchical, in which case it can be faster, or it's distributed, in which case it's socially "fairer and nicer" but slower. Neither is right; they're both choices.

AI might break this for a few reasons. In the realm of software, it's hard for one person to be the iron-fisted dictator of a team beyond a certain scale because there is just too much to know (I've called this the prima donna death spiral; I'll say more on this in the chapter on leadership). But an LLM, with a good state in a vector database, might not have this problem: It might be able to remember many more details about the project and direct at scale.

That leaves social. Again, there's an interesting effect possible with AI in that the AI can exist "outside" of social norms. We already do this without realizing it because social media makes choices in communication for us (we call this ranking). It mediates (literally) between humans, so if your friend posts something boring, you are free from the social awkwardness of politely responding (mostly) because they don't really see you seeing it (again, mostly).

It might be the case that an LLM could act as that kind of neutral arbiter in a group of coders, taking on responsibility for decisions without any one person having to feel that it's their responsibility directly, making the consensus choice more efficient. It's probably more complicated than that, but this may be the kind of effect we will see as programming teams make more use of AI,

120   Sam Schillace

not just the copilots for individuals we see now but for whole teams.

Not wrong isn't the same as right

A long time ago (way back in the '90s), at the beginning of the dot-com era, I wrote some software that became Claris HomePage. Claris was a company attached to Apple that built desktop applications.

The internet was just taking off at the time. It was very clear that it was like the current moment (April 2023) in some ways: There was obviously a lot of disruption, economic opportunity, and things to be built for a new paradigm. Claris very much wanted to have an entry, as did everyone else at the time.

Claris looked at the new paradigm through the lens of what it knew. It was a desktop app company building desktop software (as was pretty much everyone at that moment; the internet had really just barely started), so they looked at the internet, decided web pages were the central and important feature, and decided the right thing to do was to build a desktop web page creator/editor.

This wasn't wrong, exactly. It had value, and people did use it. But it wasn't really very right either. It wasn't eBay, Craigslist, Google, Amazon, Facebook, or any of several other attempts to understand the new model more natively. It was a footnote, but you couldn't point at it at the moment and say that it was "wrong." It just wasn't very "right."

This is happening again with AI, no doubt. There are lots of places where someone is adding a bit of AI to an existing application or workflow. They're better — again, not completely wrong — but they're not fully transformative either.

It's hard to be in the middle of a paradigm shift like this. It's stressful and uncertain to find new things to do, so once you've found one thing that makes sense, it's easy to want to attach to it and not look further. But that strategy can easily lead you to do something that isn't wrong but isn't right either.

Be a centaur
How to ride the exponential wave

It's been a year (in December 2023) since ChatGPT was launched. As with all disruptive things, it's had unexpected impacts as well as disappointments where it wasn't as magical as we wanted it to be. We are, as always, becoming accustomed to it, so even though a lot is still changing in the world of AI, there is a temptation to settle into the "new normal."

But that's a mistake. We are still very much on an exponential curve in terms of the capabilities of LLMs (and likely other kinds of models). Exponential curves are hard to understand; we tend to think linearly by default, so it's hard to really visualize something that is moving faster than that. There's an additional challenge, too: At low values (that is, in the early stages), linear curves can actually be higher than exponential curves. This is the reason for the famous adage, "Technology is overestimated in the near term, but underestimated in the long term." We have a linear model of impact, which is higher than exponential at first but much lower eventually.

So, given that a lot has changed and even more will change, what do we do? How do you ride this wave? The best thing to do, other than just trying your best to stay familiar and educated with what is happening, is to look for invariants. What isn't likely to change, or will change in degree but not nature, as this space evolves?

I can think of some things. Models will continue to get smarter and fill in the gaps they have now. Things like planning, hallucinations, token windows, performance, and cost will all continue to get better. It's hard to find a new thing to do — the "0 to 1" problem — but it's much easier to apply a lot of effort in parallel to optimize it once you've found it. So anything that can be incrementally improved (even if that increment is very expensive in capital) is likely to be.

The landscape of models will also continue to get more complex. There will be lots of choices for developers. Some will be at the frontier stage, but more and more, they will be looking at niches, places where some specific mix of data, cost, latency, and quality is better served by a different approach. This will continue to drive the need for tools to manage all of this.

We are likely to continue to spend more and more time in front of these models, probably in some kind of "assistant" interface. It's likely that ChatGPT isn't the end state of that and that we will continue to get richer multimodal interactions: voice, image, video, and gesture. If this goes far enough, we will get models that can generate a user interface on the fly (and perhaps consume it as well, operating GUI applications for the user). Even if we don't get that far, it's very clear that users will spend more and more time starting with and working with a model, and if your business doesn't accommodate this, it's likely to be disrupted.

Finally, what do we do, as people, to stay relevant and valuable? The answer is complex, and we aren't going to know all of it just yet (Just imagine explaining the soft-

124   Sam Schillace

ware business to a farmer in 1860, telling them, "Don't worry, farms will be automated, there will be only a few percent as many farm jobs as there are now, but your kids and grandkids will have plenty to do."). But one answer probably comes from the world of chess, where "centaurs," teams composed of humans in conjunction with models, can be very competitive. In fact, in that world, there are humans who are fairly weak chess players on their own but who are very good at managing the model's ability to explore and who are very effective as a result.

This seems to be a constant: There will always be some advantage or value in being able to work with a model. In that world, the premium is not on learning facts (which can be looked up or explained by the model) but on learning reasoning. How does the world work? What is Occam's razor? Is it a falsifiable hypothesis or a Fermi question? How do you explore a space you don't know anything about without being fooled or lost? What are the basic behaviors of physics, societies, laws, and politics?

We are moving from a world where the value isn't as much in being able to answer the question but in being able to ask the right one. Be a centaur! As the models get better, you will too.

From heuristics to precision
What is in reach now?

It's well known that networking (with other people, not just computers) is important. All of the jobs in the last half of my career come from personal connections somehow, for example. And we all know that it's important to have good networks and also to watch for bias in how we form and use them.

Why are networks so important? Well, they're a great filtering heuristic for one reason. I am not the best <programmer, creative person, engineering leader, whatever> in the world, but I'm a known quantity to my friends. They have a good model in their minds of what the job might be and a good model of me, so they can perform a fairly high-fidelity matchup. That's much easier than looking through lots of messy data in the real world.

But LLMs are pretty good at looking at messy data, and they don't get tired. So, it's not super hard to imagine them getting involved in hiring decisions, at least at the level of filtering resumes. But we probably don't want them to just do dumb, language-level filtering. We want them to have a rich model of what we need and to build a rich model of each person. This might not seem practical now, but things I've seen and worked on have made me believe it's entirely possible soon.

So now we have a world where we can have a new kind of precision at scale. We already have precision at scale when the data is regular (even if it's just a little regular, like credit card transactions, we can do things like spot fraud at scale). The boundary of that precision is exactly where we have trouble ingesting and managing data at scale because of some irregularity or complexity. But LLMs are pretty good at extracting meaning and intent from natural language and other irregular forms. So that boundary is going to move.

What will that mean? Different things in different contexts, no doubt. There will probably be places we don't like the additional precision and control (like an employer reading your emails for "tone" or "politeness"), but other domains where we will like it (Maybe it makes hiring both better and fairer by enabling employers to look more deeply and thoughtfully at candidates. And maybe it makes employment better by helping you find a job that's a perfect fit.).

In some sense, computing is always about scaling up our ability to think in some particular way. LLMs give us some new ways in which we can scale that thinking. We've likely just begun to see the impacts of that.

A new kind of coding

I've been working on building code around LLMs since GPT-4 came out. Not, "Here's a handcrafted prompt that does a cool party trick," but instead really confronting what it takes to build real, repeatable, robust programs on top of these models, using multiple inference calls to multiple models. They have interesting properties, limits, and capabilities.

Code is very good at dealing with process, precision, and syntax. It's repeatable and precise but brittle. LLMs are the converse — not repeatable (unless you set the temperature to 0, they are stochastic), and not precise — but they are very broad, very flexible, and starting to be capable of dealing with semantics: meaning and goals.

Building programs out of these pieces is a little like building a cyborg; you need both kinds of pieces, and the boundary between them is hard to get right. It's important to think about what the code is good for versus what the models are good for and partition tasks accordingly.

For example, I've been working on generating very long documents, like textbooks, that have more than 60,000 words in them. The models are very good at generating content in small pieces but less so at keeping track of all of the pieces and making sure they all get done — that's a good job for code.

One other pattern that is emerging is that it's actually kind of helpful to think about "how would I do this task" as a precursor to designing a program. Often, what is hard for the person is hard for the model, and vice versa. So, trying to get the model to do something all at once in some rigid way rarely works well. Breaking the task down into smaller pieces and letting the model "reason through it" works better.

I'll be continuing to share tools and ideas and code. This moment feels to me, as an engineer, a lot like the early web, where the development patterns were similar but not identical. Some desktop techniques were useful in, for example, building a web page, but for building fully scalable web-native applications, we had to learn which things worked and which didn't in the new context.

I believe we will all spend the next few years repeating this process, but with AI. As model capabilities emerge and develop, we will begin learning the right patterns and best practices. It's important right now to be open-minded and curious, and to read and experiment a lot. There will be lots of false starts and funny ideas in the early stages — and that's okay!

No Prize for Pessimism   129

Build unreasonable things
Lessons from the last epoch

We are clearly at the beginning of a new tech epoch with AI. In many ways, the moment feels like the early internet — we can see a vast landscape of possibilities opening up, entirely new categories of problems that can be solved, but in the near term, there are all kinds of implementation challenges.

While it's important (and necessary) to solve those problems, it's even more important not to be stuck to them. It's very easy to build something "small" that feels satisfying, just because it's new and possible at all. We are seeing much of that right now, such as "dancing bears" and party tricks that aren't scalable or useful.

The trick at this moment is to be ambitious to the point that your friends think you're kind of nuts. It might be hard to imagine if you grew up with them, but there was a point in time when "the infinite bookstore" (Amazon) or "all of the world's knowledge" sounded like a crazy, impossible dream. There used to be debates about how expensive search was, that it was an impractical idea — and lots of skepticism about e-commerce ever being widely adopted.

Right now, advanced AI models are hard to get access to, expensive, and slow to call. That will all change. Once an idea has been discovered, which is hard, optimizing it for speed and cost is something that is both easier to do (so more people can help with it) and parallelizable, so it can happen much more quickly. We should expect that here.

The right thing to do is to pretend the performance and cost are already where they need to be. This is one of the "What if?" questions. What would you build once the LLMs are a few more generations better and only take 10 milliseconds to call? What if you had something two or three generations smarter that you could call not just once per user interaction but thousands of times (which is how most large services work today — a web search or product page load is thousands of internal RPCs in parallel)?

It's hard to do this; it's quite a leap of faith, and we often don't get it right. But it's essential when we are on an accelerating curve like this: You can't just aim for what's next; you have to be unreasonable and aim for what could become.

In praise of hard optimization
"Okay, Grampa, tell us again how it was hard in the olden days?"

There is understandable excitement about new AI models like ChatGPT, GPT-3, and others. There are lots of really powerful new things that can be done with them, and we are almost certainly going to spend the next years figuring out new coding techniques and best practices to build with them.

One striking thing, though, is that we are back in a land of scarcity in a way that I haven't seen in a while. At the start of my career (way back in the '80s), programmers had to spend a lot of time worrying about memory, CPU performance, etc. We still do, but the systems we work with now have much more flexibility and, often, scarcity can be solved with dollars. You're not worrying so much about whether you can get something to work at all; you're worrying about whether your cloud bill will be too high. And some of these problems, like graphics performance, have been solved for decades by dedicated chips. We rarely have hard performance constraints unless we also have hard financial ones.

But in the earlier phases of my career, heroics were often needed to even get something working at all. Thirty-three years ago, a friend of mine and I built a video game that ran on the Mac Classic. Tiny screen (512 x 342 pixels, less than a square inch on an iPhone!), the game was monochrome, and we **still** struggled to get the frame rate up to 16 frames per second, the bare minimum to make it usable. We wound up having to do things like

132   Sam Schillace

run-length encoding of the frame buffers and hidden line removal so there were fewer pixels to think about and draw. Flash forward to 2006, writing JavaScript for the early versions of Writely, when the browser runtimes were very limited. There was not much memory and a limited number of cycles per request, so we had to be very careful with what we did on the client. We let the browser do most of the work and only did things like diffing in the most basic and optimized way.

I see this in AI today in many ways. There are time constraints: Larger models take a long time to run, so you can't use them for anything in real time. There are space constraints: LLMs have token window limits, so you have a small amount of data you can pass in to get a result. And there is hardware scarcity: GPUs are in demand, and there is general pressure on chip supply, so it's hard to build a large service or use arbitrarily large amounts of one.

All of this points back to a kind of engineering that may not be familiar to folks new to coding (note how old my examples have to be!). It's easy to look at a set of constraints like that and think, "Well, it can't be done right now; I'll just wait for the world to solve it for me," because, for a long time, that has been the right approach; speed to market mattered more, and scale was mostly

No Prize for Pessimism   133

a matter of dollars. Those dollars used to be cheap, but now they're not, and the things they can buy (non-GPU servers, hard disks) aren't as interesting to the next wave of problems. It's time to optimize!

Almost every major technical transition (and we are for sure in one with AI now) comes with this kind of challenge in the early phases. Tools are weak and ill-suited to new tasks, performance is a challenge on many dimensions, and optimization techniques may be non-existent or not well known. It's a great time to dust off your CS fundamentals and do heavy lifting. This is a mindset shift if you aren't used to it, but for many of us, it's a flashback to how it used to be regularly. It's not fun, but it is satisfying when you find your way through what looks like an impossible thicket.

As we go through this transition, it's important to not take no for an answer when confronted with this kind of problem. Our whole job as engineers is to find our way through these and, at this moment, those skills are more valuable than ever.

What's the next development model?

In my lifetime, I have encountered several computing models: mainframe, desktop, web, mobile, cloud … and now AI.

Roughly speaking, each one generates some kind of programming paradigm that solves a dominant problem in that generation. Mainframes started with assembly but did a lot of work on compilers and languages as programs got more complex and too hard to manage. Desktops were tightly bound to physical media, so ship cycles mattered, and we got waterfall programming. We lost that constraint, so we got agile — quality got loose for a while and then started to get automated, so when mobile/cloud showed up, we were ready for full CI/CD and more automated dev lifecycles. Monitoring, debugging, refactoring, environments, and code reuse (OSS) all evolved alongside them.

AI seems like another shift. Not just AI in the sense of building/training/hosting a model. That of course has its own set of tools and workflows. Here I'm talking more about the application level. What are the patterns there? We've seen a little bit of it already with things like prompt engineering, but that feels transient to me.

I can see some obvious challenges. We are building things with very open-ended behavior, so monitoring probably needs to be at the same level as semantic telemetry. Is the agent being nice and friendly? Is it stuck on a step of a task? How do you do regression testing as base models change? How do you decide where to run an inference, in real-time, for the right cost/quality tradeoff?

At a higher level, I wonder what the right strategy for code interoperability and reuse is. Is it all just natural language? That seems inefficient, but falling back to a rigid schema seems to lose some of the points and power of LLMs. What's the equivalent of a package or gem in the world of LLMs? Do we package memory and expertise for reuse? Is that a pipeline, or do we merge memories or weights directly into models? How do we test and manage all of this?

There's also some tantalizing re-entrance. Models understand code; if we give them full access to all of the dev tools, how much can they manage? Is the next step past serverless "codeless" or something like that, where the LLM is a full partner, and teams abstract up another level and mostly work on intent? Or maybe our dev tools just get much smarter, debugging becomes "This is doing X, please figure out why that is," and telemetry becomes "Tell me what users don't like," etc.

I don't know if it's possible to spot this ahead of time from first principles. I remember how controversial even agile was at first. What mostly happens is lots of teams try lots of things, and the best techniques emerge, like natural selection.

Whatever the answer, it seems pretty clear that the practice of development is about to radically change again. Another step up the ladder!

New paradigms need new tools
And it takes some time

There have been four major paradigms in programming in the history of the computer industry: mainframe, PC, internet, and mobile. AI is almost certainly the next one. Each new paradigm presents new problems and needs new engineering approaches, not just new tools but new systems and ways of thinking.

Mainframes were big enough that you could start to write programs that were too hard to keep in one person's head, and they were big enough to have more than one user, so we invented things like compilers, multitasking operating systems, and all of the tools that went with them. PCs were smaller and lighter; we didn't need the same kind of user management (mostly), but we did need to manage fleets of machines, secure them, and build more complex (and eventually graphical) applications, which needed things like application frameworks, automated testing, SQL databases, and local file systems, etc.

The internet brought challenges around scale: Suddenly, we had to manage applications spread across

No Prize for Pessimism   137

many machines with different kinds of UI surfaces, like the browser. This meant we had to learn how to do things like orchestration on the server (distributed systems), larger databases (NoSQL), app frameworks that combined the server and client side, and programming disciplines like CI/CD and automated testing. Mobile might be thought of as an extension of this paradigm, but it brought new things forward, like deeper network integration, always-on telemetry, etc.

So now we are in the age of AI. What tools do we need here? Well, as app developers (not folks who create AI models, but who consume them), we need to perform many of the same tasks as before: build pipelines, QA and regression automation, load optimization and scheduling, user data management and security, and remote telemetry. These all have to be managed in the context of something that is both stochastic and semantic; often, the things we need to do telemetry testing or scheduling on are in the semantic realm. For example, we might want to know if one of our prompts still produces an answer that is "nice" in a new generation of a base model, and we can't do byte-wise comparisons because the models are stochastic. Instead of monitoring latency, we need to monitor and manage something more in the domain of meaning. This becomes very important as we build more independent agents.

There are also new problems that emerge. We need to be able to decide where to run a given inference; we might not always want to run it remotely, or we might want to run it on a cheaper model. So we need some way to do "scheduling" where we can predict the quality of an inference on a given model, without actually running it first.

There will be challenges and new approaches to all of these problems. Sometimes, the old tools will work fine (like compilers) or evolve (like interpreters). Sometimes, we will have to rethink the overall approach (like the shift from waterfall to agile that the internet both enabled and encouraged). There are lots of new problems to think about, too; how do you prevent and defend against prompt injection attacks, for example? (This reminds me of us all learning to deal with SQL injection and XSS attacks.)

This process won't happen overnight. Each of the large shifts mentioned previously took years to work out properly. There will be false starts, religious technical wars, ideas that sound good but are impractical, etc. It's quite possible that the approaches that work best aren't obvious at all yet because no one has built a complex or ambitious enough product to need them.

The right thing to do is to be open to change. There will be new approaches that feel wrong at first. It's valuable to ground ourselves in what works, solve specific problems, and do as much of this as possible in the open and together so we can all learn and debug in parallel. Don't be tempted to write off early attempts because they aren't perfect. It's much better to be pragmatic and think "What if?" about new tools and approaches — and to help solve problems rather than throw rocks at the people working on them.

The developers and companies who were most successful in the prior transitions were universally the ones who had this approach to the new paradigm, both its problems and potential.

Finding the way in

One key challenge in building products is getting past incumbents. Sometimes, you can build a very valuable new product but be blocked in some way by a larger established player — whether through regulation, control of distribution, being outspent, or sometimes even just having someone else add a weaker version of your product that prevents users from adopting yours.

Way back when we did Writely, we had a bit of this dilemma. Microsoft (for whom I work now!) has a huge mindshare in the enterprise world and lots of credibility. It was very hard to convince a CIO to take a chance on a startup (or, later, Google). We weren't blocked, but we didn't have a good way to compete directly with them.

So, we made use of the "innovators dilemma": We looked for users (like students) the incumbent wasn't focused on, or who seemed less valuable or harder to access to them, but for whom our solution worked really well. That opened the door. Small businesses, education, and global customers were all early adopters that saw value and got us going.

In the world of AI now, the same thing is likely to be true. There are lots of great ideas out there, but it may not be possible to take them through the front door. In teaching, law, and medicine, for example, there are lots of underserved communities and markets that will be easier to approach than the mainstream — look for them.

Incumbents always have many advantages. However, one of the advantages of being a startup is that you can fit into small spaces and understand obscure use cases, and sometimes these grow into really big opportunities.

Where is the autonomy in AI?
When do we get to take our hands off the wheel?

I'm not a deep ML expert. I understand the science pretty well, and I read a lot of papers, but I'm more of a builder of applications than of models (I like to say I consume them, not create them). I've been working with LLM-based systems for about a year now and closely watching both my own and everyone else's work. And I'm struck by something: No matter how capable LLM-based solutions are, there always has to be a "human in the loop." That is, they can do great things, but only if we hold their hands in certain ways.

Some of this is obvious, and some isn't. Prompting strategies like "chain of thought" is a subtle way that we are doing some of the work for the model; we are doing the "metacognition" of picking a cognitive strategy that is likely to suit the problem first. Then the model "bakes the recipe" very well, but only once we've picked it.

For longer tasks, the models can do each step, but we need to do error correction, goal maintenance, and completeness evaluation. I can do something fixed, like the "textbook factory" experiment I built that takes a sentence of input and uses about 200 lines of Python 5 prompts and 600 calls to the model to build out a full-year textbook. That works well enough (mod hallucinations), but it's only good for that one task and I had to — again — do the metacognitive organization of writing the program.

Or I can walk something like ChatGPT through a complex process like writing a book or a program, but I

142   Sam Schillace

have to do the walking. I have to prompt at each stage, correct it when it's wrong, curate the answers, and decide what the next task is and when we are done. You can't just say "write me a book" and then "mmm hmm ... keep going" over and over and get a good result.

Think of a task like "write a clone of the vi text editor in Python from scratch." You could give that to an intern or CS grad with a simple, closed code environment and expect that they would spend some substantial time on it and get it at least most of the way done. We can easily imagine the metacognitive planning we'd do, too, if we were that intern: Start by organizing requirements, think about modules (editor, event handler, macros, testing), iterate and evaluate the design against the spec, recurse down to smaller and smaller pieces of each task (write module 1 - first step name it, second step define classes, etc.), until we're done.

It seems hard to build that strategy into a model or LLM-based solution. I might be unaware, but I don't think anyone has done something like the above: Give a long, complex arbitrary programming task that doesn't fit into a single token context (or even 100 of them) and some system can complete it without intervention in some simple environment. Much less a more complex system that can successfully navigate the real world for a complex, long-running task, like "Here are my data sources and last quarter's results, re-create this set of artifacts with the current quarter's results" or "Look through our org chart and give me a detailed salary analysis relative to local markets, highlighting any more than one sigma outliers." Or, especially, open-ended or underdefined problems where some questions need to be asked ("all of the features? Which version of vi?") along the way.

I don't know why this is. Some would argue the models aren't smart enough yet. Some might say that we don't yet have the right architecture on top of them (I fairly strongly suspect there is some kind of generic "recursive hierarchical problem-solving strategy" that we haven't found like people had to figure out B-trees and such way back in the day). It's hard to describe metacognition, but it's not impossible. Maybe the errors from stochasticity compound as you go up the levels (which might make sense as the leverage increases), and there's some subtle error correction approach we haven't found yet. It's definitely the case that chat isn't the best UI metaphor, and we need better ways to have messy, branched, asynchronous iteration more akin to human workflows and conversations.

Setting aside the question of whether it's a good idea to do this or not, it's very striking to me that, after a year of more or less everyone trying very hard, we haven't gotten very far with autonomy, even in "toy" environments like pure coding. I don't know what it means, but it's hard not to notice.

A crisis in epistemology
What should we teach in the age of AI?

One of the questions being asked a lot these days is, "What is the right thing to teach (or learn) given the capabilities of AI?" It's a valid question. We teach arithmetic even though few of us do a lot of it manually now, but it's still important to know how numbers work. What's the line now that AI can do so much more for us? How much programming should you learn versus just learning to get a model to do it?

The easy answer to this tends to be something like, "We happen to be teaching precisely the correct fundamentals to have a complete understanding of the world. Yeah, liberal arts education! Just do that." I think there is some truth to that, but it's more complex than that.

Epistemology is the study of knowing (roughly). How do we know what we know? There are (in my layman's sense) roughly three ways to "know" something: experience it personally, hear it from a trusted source (expert), or believe a large enough crowd of non-experts.

Each of these involves some form of trust: in your senses, in the expert or system that produced the information, or in statistics.

The internet was already breaking all of these mechanisms, and AI will break them further. We will have better and better fake evidence, so you have to be careful about primary observation. There are better and better fake experts (the famous argument about AI being good at definitive bullshit). Social media already gave us crowds that believe pretty much anything you want (Hello, flat Earthers, I hear you're all around the globe!), and AI will give us that even more strongly: assistants that potentially tell us whatever we want to hear in elaborate, patient detail.

So what is the skillset to teach? It has to be "the ability to learn well." What does that mean? Well, it's a mixture of things: Some fundamentals about how the world works, to start. Things like Occam's razor or the idea of a falsifiable hypothesis, the different forms of cognitive bias and error, the fundamentals of statistics and probability, etc. And some specifics about science, math, biology, art, politics, and history.

The goal is to be able to come into a complex environment with contradictory opinions but good resources (like search and AI) and be able to rigorously and accurately form a good opinion about a topic you are formerly unfamiliar with. This is useful not just for navigating the world but for navigating tasks — it's likely that work will largely become that kind of abstract behavior, where you will have to use the AI to help navigate something unfamiliar because, by definition, anything familiar will become rote and automated soon.

This is as hard as any educational process ever is because there are always opinions about what "truth" is and what is worth learning. For myself, I look to pragmatism: What works? I like the scientific method. The science community has all kinds of problems and perverse incentives, but it is as robust and rigorous as they can make it; it's constantly trying to improve; it is centered on doubt and asking questions rather than certainty and making assertions; and the basic patterns of hypothesize, test, control, observe, and iterate have gotten us pretty far.

The things we have needed to learn have changed many times throughout history. Until fairly recently, most of what you needed to know was how to feed yourself or the basics of your trade. Then, the world opened up, and we began teaching skills that applied more broadly to the industrial age. We are on the cusp of another mass shift. The Industrial Revolution was the first time we had a massive surplus of physical energy available as a species; this is the first time we've had a massive surplus of cognitive energy available. What we will need to learn and be able to do will shift accordingly, just like it has in the past.

The future of work

Learn to play

I was asked to speak at a conference for data analysts. Of course, the big question on their minds was "What happens to our jobs?" followed by "What should we learn/teach?"

These are good questions, and many different specialties are asking them. I always reach back to the idea that, 150 years ago, 2% of the world population **wasn't** working in agriculture, and now 2% **are**. That's a huge shift! It would almost certainly have been impossible to explain to a farm worker back then what jobs would look like now — so much of what we do is built up on other capabilities and needs.

But thinking about data analysis specifically, it occurred to me that I've never had a team say "Gee, we have so much analytical capability, we don't know what to do with it!" In fact, the opposite is of course always true: We always want more. And analysis is kind of funny in that questions beget more questions; as a manager, you often have to hold back because the team can only do so much.

But now we are in a world where "everyone" can do some aspect of this role. That's great! Cheap experiments are the heart of innovation (and cheap questions are the same thing; every experiment is just a question in a different form). Why? Because the good ideas are usually surprising and counterintuitive. If they're also costly to explore, it makes it very hard to look at them. Cheap, easy, fast experiments are vital. So, it's great that we have more capacity now.

So what do you learn, and what's the value of professional analysts? Well, think about the most basic bit of statistical understanding: mean versus median. If you don't understand this, you can't ask meaningful questions. "Look at that room! Everyone is a billionaire on average!" (Actually, it's Bill Gates and five other random people; you want to look at the median, not the mean).

So even though the tools are more available, and the mechanics are easier to drive (so things like R are less valuable to memorize), knowing what to ask and how to understand it is going to be more and more important, not just in this field, but in all of them.

Chapter 6

Product Design

The thermodynamics of user laziness
How adoption really works

One of the things I have said and believe is that users (that is, all of us) are **lazy**. That's a good thing! As makers of all stripes, we love what we make and think other people will. But the reality is that we are all busy and don't have time for something that makes our lives a little bit better. We're "lazy." This isn't just in software — any product that involves a user choice has to reckon with this.

But what does that mean, precisely? What's better? One thing that's been commonly said is something either has to be totally new (that is, solve a job that has no solution at all) or be 10 times better. Why?

I look to thermodynamics for this, specifically the second law (stick with me!) that says that entropy always wins. In a closed system, entropy can only go up — things become more disordered. Okay, but why, then, do we have order at all? If a cup of hot coffee can only cool down (in a room temperature room), how is it hot ever? Why doesn't that violate the law of thermodynamics?

Because of that definition of closed system. If you look far enough out — say, to include the solar system — entropy in that system as a whole is increasing. The sun is increasing entropy massively as it burns fuel, which sends some heat here, which ultimately winds up heating that cup of coffee. All along the way, bits of entropy are increasing to deliver that last small payload of order.

This is how it is with user laziness. Companies will design a new product that's a little bit better, and then wonder why people don't use it. This is better! Why aren't they using it more?

The answer is that closed system again: Users are lazy in the totality of their lives. If the effort to move to a new habit plus the savings of the new habit isn't a net win in terms of overall effort, they won't do it. The hassle of learning something new and remembering to change a default behavior ends up being more work and thus never happens. It's only when there is so much novel value that the effort of adoption is swamped by that value (10 times better or a solution where none existed) that users will move.

This seems obvious, but it happens over and over again. Throw a rock, and you can see this in almost all of the work being done with generative AI right now. There are some genuinely good and transformative things being done, and some of those will be successful. But there are also many, many incremental improvements being made in the hope of changing user behavior, which is very unlikely to work, even within an app, much less across it. Even really disruptive things like image generation won't work if the additional work asked of the user still nets out to a negative return on time.

It's hard to do because we love what we build, but it's always important for a maker to take in the thermodynamic totality of the user, and ask, in an honest way, whether the solution is really, truly easier for them overall. If you can't honestly answer that, no amount of wishful thinking or good intentions or so-called "funnel packing" will move them.

Trust is a design issue
Or maybe the other way around

When you build a piece of software (or present any other service), you are at some level making a set of implicit promises to the end user:

- You have thought through the problem domain carefully enough to understand it well.
- You have the skills and tools to build a solution that works for the problems described.

- You have tested and honed the solution so that it does, in fact, work the way you expect, and you can be relied on to keep the service or software healthy over a reasonable time.

All of these ask the customer to trust you, and respecting that trust is core to being successful. Different aspects of these promises are important to various degrees depending on the domain. Sometimes these are really obvious: Reliability might be very important in some domains, or data privacy might be important in another. These tend to be the things we focus on the most; if you write bank software and it can't add accurately, you know pretty fast that you've breached trust.

However, there are more subtle promises that get overlooked, like user experience and latency. It's very easy for teams to focus on adding new functionality (We're makers! We like making things! Removing them, not so much) at the expense of overall complexity (tech debt) or performance (The toolbar is slow to load but look, so many buttons!). This is a very common pattern that teams fall into. Software, over time, makes compromises as it grows, and teams slowly get comfortable with the uncomfortable feeling of letting their users down. Great teams don't do this; they recognize that every new feature involves a tradeoff, and they're careful to consider the promises made already before making new ones.

It's important to keep these other promises in mind as we build. Our sense of frustration with software that loads slowly or works poorly is often combined with a feeling of being betrayed: I used your software! Why are you breaking my heart?

Worse is better
The role of simplicity

Many years ago, a paper decried why seemingly "worse" technical languages were beating out "better ones." The author concluded that it has to do with the order in which certain properties are respected — does the language favor completeness and correctness over simplicity? Or is simplicity the most important thing, followed by correctness (of what's there) and completeness being neglected when there's a conflict?

It's a good way to look at software design as well. It's very hard for engineers to let go of parts of a problem. We tend to be obsessive types so, if it occurs to us, we want to fix it. This is especially true in the world of enterprise software, where every buyer wants every item on their checklist, so solutions tend to have a kitchen sink feel to them. It's hard to say no! It's even harder to take something out once it's in.

We need to remember that even though software, in many ways, is less constrained than physical systems, it still has constraints. We can't do everything, even if the constraint is just our ability to pay attention to and test all of it. So, if a team implicitly favors completeness, simplicity, and even sometimes correctness, it will eventually suffer. This is why really good design in software is so hard to do; most teams, most of the time, feel enormous internal and external pressure to add complexity, not take it away.

The point of software, though, is to get a job done. In the consumer space, it's very important to articulate that value clearly and simply, so "worse is better" usually prevails. Winners tend to be the fastest but also the clearest and easiest to use. But it's hard to keep working on simplicity: Isn't it simple enough? Let's add some features! Every company does this, and this drive to make things more complex and fragile over time leads to openings for competitors to come back in. This is a big part of where the "consumerization of enterprise" came from: Old-school enterprise companies had gotten overgrown and hard to use. New ones came in that were much more focused on simplicity and ease of use, and even though they were "worse" in terms of the old regime, they were better for users and successful in a world where there is more user choice for business software.

It's easy to get lost in software. We want to build large, complete, complex systems, and software is really good for that. But it's not always the right thing for the user, and sometimes, it's not even the right way to succeed. Sometimes worse really is better.

Understanding, explaining, and simplicity

One of the best ways to see if you actually understand something is to try to explain it to someone else. It very quickly exposes gaps where you thought you had a clear understanding when in reality you just had a passing one. (This is also one of the things I like about math and code; they are other ways of explaining something and seeing if you really understand it).

We value clean, simple designs in many areas: user interface, code, and even organizational structure. But we often don't get designs that are as simple as we want; why is that? Well, simplification is, again, another form of explaining things. To simplify something, you have to be able to break it down into its most fundamental, irreducible pieces and then only present exactly what's needed. This is almost exactly the process of explaining something to someone — breaking it down into just those ideas you need to understand it.

So, just as trying to explain something will expose how well you understand it, trying to *simplify* something (a design, a team, an email) will show you how well you understand the problem. This is very useful! Usually, the instinct is to add to a design until it seems to do what you want, but the real trick is to take it away until you understand it will be enough for it to be simple.

Beware of programmer traps!

If you are in the tech industry, you've probably had the experience of trying to explain something to someone in your life who isn't technical (Usually people say "to your grandmother," but I once had someone say, "But what if my grandmother was Ada Lovelace?" — good call on stereotyping there). You will find that it can be surprisingly hard to explain things that, to a programmer, seem obvious, like "What's code?" or "What's the difference between the browser, a web page, a server, and the internet?" and "Why don't the cat pictures load faster?"

This is because we have self-selected as people who are very comfortable with complex abstractions. Most of what we deal with in the world of software isn't physical. It's not even "real" in the sense most people would mean;

it's all "just" some weird math that only uses 1s and 0s, lots and lots of them. What do you mean there are "trees" in there? What's a pixel, and do you need different colored numbers for them? Because so many people who are good at this skill are in the tech industry, it's very easy to think that *everyone* is good at it — after all, all of the people you meet daily are.

This means that sometimes you can design a feature or a product that makes a lot of sense to you as a programmer or abstract thinker, and when you share it with people around you, they will nod and say "right on," but when you launch it, it will go nowhere because the mass of the world's population can't do the math in their heads to make it make sense, at least not comfortably.

This is a "programmer trap" — and these patterns seem to repeat themselves. Typically, someone decides that all of the little edge cases and irrational features in a design are irritating, and it would be easier to just make one single abstract, composable, programmable model. Wave was a good example of this, and there is quite a graveyard of people trying to do "personal information managers" or "programmable documents" or things in this general area that try to take the necessary mess of some domain and make it into an abstract model, or add a "little programming" to make it "easier." Slack is probably a good example of the opposite direction: Rather than making a very abstract and pure program ("it's JUST IRC!!"), they focused on real workflow and real user patterns and let it be idiosyncratic where it needed to be to work well. In some sense, this is "the rise of worse is better," but for products that are complex and complete, it usually doesn't beat simple and workable.

The only real cure for this is awareness. Remember that things that make sense to you might not make sense to everyone because you may have unique skills and live with others of similar perspectives. When programmers hear "You can write code to make it work," they are excited, but the rest of the world is either terrified or confused. It's a very good reminder that user testing, even in the form of "Explain this to the next person I randomly meet who's not a tech person," is very valuable.

Look out for the traps!

The abstraction trap

Abstraction — the generalization of a single instance into a repeatable, sometimes parametrizable pattern that can be reasoned about — is the bread and butter of the technical world. Think briefly about the levels of abstraction that needed to be climbed for you to read words on a website. Every computer in the world fundamentally only does one thing: binary math! That's it. But we use abstraction to turn that math into letters stored in a database, bits of an internet packet sent to a server and encoded on a magnetic medium somewhere, and, finally, even small colored dots on a screen that look like the letters we want.

You read that paragraph and probably thought, "Duh!" It is pretty obvious ... to us. And that's really the point of this letter. If you've selected into and been successful in the tech industry at all, and most particularly if you work closely with software, you have a very above-average ability to consume, internalize, and use abstraction. You're likely very comfortable with it.

But most people aren't. How often have you heard someone say, "Oh, I just don't get computers at all," and you laugh a bit at them because it just doesn't seem that hard to you? We've all had moments where we don't really understand why something is so hard for someone to understand; almost always, it's about some kind of abstraction that's hard to handle.

This can leak over into our software design very easily, and it's related to the "programmer trap" we just discussed. It's not enough to test software on people in the industry who can not only parse abstraction but

No Prize for Pessimism 161

who are also insiders who understand common design patterns (and IT professionals are "in the industry"). Often, when I get stuck with a piece of software, I can think, "Okay, what was the programmer probably thinking when they built this?" and it can help me understand. Most people outside of the industry can't do that. They need something that has a more familiar narrative flow and asks less of them in terms of complex mental models.

If you like, you could think of this as a form of accessibility design. If we were designing software that was very difficult to use if you were colorblind (which we sometimes do inadvertently), the correct response wouldn't be "learn color" any more than the correct response here is "learn abstraction."

The right answer is always to meet the user's needs. This is easy to do when we can understand and feel those needs ourselves. It's much harder when we have our own biases and blind spots that are more common in the industry — and in that case, we have to take extra care to design for people who are less comfortable with abstraction than we happen to be. This means lots of user feedback, user experience research, constantly revising rough edges, good telemetry to see where people get lost, and all the usual suspects.

Fit then optimize
Not the other way around

Getting started on a new software project or company is complex. Choices you make early on can be felt throughout the project's lifetime, so it's important to get things right. At the same time, it's important to get moving and solve a useful problem — to get product/market fit for a company. Often, teams and engineers struggle with the balance between these two goals.

One common pattern is to try to optimize early. In the early stages of any project, you're not likely to understand yet what's truly important. But there are lots of apparently interesting problems that you can work on, so it's tempting to pick one and try to optimize it. It's much easier, though, to tune something that has found a fit (or, for an internal project, is working) than it is to make something that is a highly optimized fit (or work). So, in general, it's "fit first, then optimize."

However, there's a downside to that approach that leads to a very common pattern: the hairball. One of the most critical parts of a company or project is understanding when fit is "good enough" and beginning the process of cleanup and tuning. The reality is that this is never perfect, so all successful projects have some degree of ongoing cleanup, but the bad case is that this gets neglected for too long and the project (or company) has to fully pause to do a cleanup.

This is more common than you think; many big tech companies have had an event like this in their history. There's never really a good time to shift back to architecture or optimization; once the project is rolling, it's rolling. There's no hard and fast rule for when to start the process of cleanup, but as soon as you begin getting traction, it's worth starting to observe the rough spots and forming hypotheses. Ramping up gradually here is also helpful: Don't re-architect the whole thing at once; that's no better than letting it fester. Rather, as use cases and patterns emerge, try small changes and cleanups to streamline them and see if they're repeatable. I have a mantra I use here: "One, two, platform". What does that mean? Don't build a platform until you've built at least a few instances of something that it would support. You can usually sense there is a need for one, but it's good to have a few examples before codifying the patterns, so you know you really understand them fully.

A related pattern is "We don't have fit, so let's design something that can cover all cases in our area of interest while we wait for inspiration." This leads to the opposite kind of failure, a system that is so broad that it's not useful for anything. It's good to not be optimizing early,

164   Sam Schillace

but under-optimizing is bad, too. The goal is to find a tight fit between the code you build and the problem you are solving. The ideal (never achieved) is that you never write a line of code that you don't need. Building a big general framework before you understand your problem or customer well is just asking for a different kind of waste.

So ... find fit first! It is always the best thing to do to optimize slowly ... but.

The value of iteration

A rocket launch is a highly complex thing. The launch sequence has to be thought out precisely, with every action planned and executed carefully. Every contingency has to be accounted for, and anything that can't be accounted for will block the launch or worse. In contrast, driving a car to the store is a fairly casual event: Get in the car, look out the windows, and steer it to the store, reacting to small changes along the way.

Software started with the rocket model, otherwise known as "waterfall," because at the end of the process, there was, in fact, a "launch" where the software was printed to hard media. When the cloud arrived, this method was replaced with things like agile, Kanban, and others that take a fundamentally iterative approach to not just building software but to understanding user needs. Instead of intensive research, there is intensive iteration.

But the echoes of the old mode are still with us. Often, applications ask users to perform a task or enter data in a way that is very rigid and linear. Even though the application is developed iteratively, there is little iteration in the interaction of the user and the code. Instead, programmers and product designers do their best to interpret all possible ways of interacting, and deal with them upfront, like a rocket launch.

Large language models that are now emerging might give us a chance to bring iteration back to software experiences. Think about working with a new person; you'd never give them a complete set of instructions once and then never speak again. You'd give them an approximation, watch them perform some tasks, correct, and iterate. Over time, you'd build trust together.

That sense of iteration, conversation, and trust is largely missing from applications that are built today. But it's a much more natural way to not only build user experiences but to build scale; instead of an application being one experience for millions of users, it should become a different experience for each of them — one that's fine-tuned to their needs.

This is a huge mindset and tooling shift, and it won't happen overnight. But just as the internet brought forward a huge shift in not only how software was developed, but what it was capable of (there weren't social networks, ad networks, or e-commerce before computers were widely connected), the availability of LLMs and other ML techniques has the potential to shift how we think of human/computer interactions.

Spend some time watching your programs. Look for the ways in which they make you do work instead of working for you. Think about how you would explain your patterns to those programs if you could. You'll be surprised by what you find.

Chapter 7

Humility in Leadership ... and the Prima Donna Death Spiral

I became a leader somewhat by mistake. I went to Google as part of an acquisition and had a really fun project to work on that a lot of other people were interested in. Before I knew it, there was a large team of people following me, Google promoted me to director, and that was that.

Because of that, I never really had a lot of ego attached to the leadership role — and I still don't. I just like building things and working with people, and it comes out how it comes out.

Being somewhat accidental about leadership, though, has helped me see a certain pattern. We think leadership is about ego and "progress" and "winning," but it's really about humility. The best leaders all manifest a lot of humility.

Delegation is a good example. It's a form of trust and humility to delegate work to people around you. The reverse of this is a pattern I call the "prima donna

No Prize for Pessimism   169

death spiral." This happens with new leaders, particularly technical ones. As a team forms around them, they often make a judgment that any given task is likely to be done more quickly and better by them than by delegating (often true at first). So, they arrogantly do that task (prima donna). But that lack of humility and trust in their team makes the team that much less trustable. As the team gets bigger and more disenfranchised, the pressure on the leader increases until it's unworkable (the death spiral). If you look closely, you can see how a lack of humility — trust in others in this case — leads to that pattern.

Healthy leaders think in terms of serving their teams and making them shine, not in terms of their own ego. We'll explore some other ways this pattern manifests.

Be a brave fool: Challenge your own authority

We've all seen the character in a movie who is an obvious charlatan, but who rides into town in a cloud of credentials, references, and degrees, all of which are fake but work anyway. Or maybe we've met them in real life: the "just because I say so" professors, the bosses or executives who rely on titles to stifle dissent. This happens in science, too; new paradigms often have to wait for the "old guard" to die off before they can be accepted. "Oh, you don't have a Nobel prize? Well, I do. That quantum stuff is just silly"

One enormous occupational hazard of becoming successful is that you will believe your own press, and will stop hearing important news about the world. The title on your door is so impressive! How can this entry-level engineer be right about a giant risk in your code? You'd know, wouldn't you? After all, you have the title!

Even worse than believing that about yourself is convincing others around you to believe it too. This results in some very bad effects — you won't get good input, or any input at all. Your organization gets more brittle and blind spots take on the force of legislation.

One cure for this is to be very deliberately self-deprecating. I have a bit of a potty mouth — not the best habit, but it helps signal that I don't take myself too seriously. Not being afraid to make fun of yourself ("I can't believe how dumb that was") or correct yourself in public is good too. Go out of your way to thank people who push back and give feedback. Don't be afraid to show vulnerability.

Make it clear that you are happy to debate things on technical terms. One of the smartest, most successful engineers I know will debate almost anything with almost anyone, on purely technical terms, with no negative consequences as long as the debate is in good faith.

It's very tempting to get by based on past performance or external credentials. But it's much healthier to try to always stand on the value of your work alone. It's scary — you might have a bad day, everyone does — but in the end, the closer you can get to that authentic place, the more effective you can be as a leader.

The inverted org chart

As I've said, it's easy to think about leadership from the perspective of ego. We use lots of language like "rise," "top," "achieve," etc. This shows up in organizational leadership certainly, but it also appears in other kinds of leadership: being in a conversation, working with a team on a technical design, etc. It's very easy to get your ego involved with an idea, or with your status or title, and it's easy to see this as a natural part of leadership.

But there's a different approach to leadership, sometimes referred to as "service leadership," that's much more effective. One way to put yourself in this mindset is to think about org charts. We usually talk about a single person being at the "top" of the org chart; it seems good and exalted, and something you'd strive for. But think about an "inverted" org chart for a minute, where

172   Sam Schillace

the leader is at the bottom and the people who do all the work are at the top.

This framing shows that the real purpose of the leader is to support the team. If you have this mindset, making your organization larger doesn't feel like some egotistical achievement — the pyramid is taller! — but rather an increased obligation to the larger team you now have to support.

Many other facets of leadership are also more effectively viewed through the prism of humility and service. The "prima donna death spiral" I've talked about comes from poor tradeoffs between long- and short-term team needs — and these tradeoffs are usually rooted in a sense of ego and importance. Delegation and trust are a form of anti-ego and humility, and they make teams and leaders much more effective if done well.

So … lead, for sure, but try to do it from a position of humility. It's about service, not about being elevated.

Bad communication is (almost) always your fault

One of the biggest challenges in engineering teams is making decisions. Many readers have experienced nerd wars, two engineers (or groups of engineers) at loggerheads about some topic. Why does this seem to happen so often? There are many root causes: different properties of the system can be valued differently, differences in taste or coding preference, and even differences in understanding. But these all ultimately boil down to differences in communication.

It's very easy to be frustrated with someone when they aren't understanding what you are trying to convey. After all, it's perfectly clear, isn't it? So they're just being stubborn, or mean, or dumb (we know not to assume bad intent right?).

But, of course, what you're trying to convey is clear — to you! You have it in your head already. Your job isn't to convince the other person that you have it in your head, it's to get it into their head so you can have a real conversation about it.

If you assume good intent, and you know that the burden is on you to get the idea across, then any failure of communication is your fault. Not really (and not always), but it's a good mental posture to start from. If the other person doesn't understand, you may have to do some digging to uncover what's blocking them. Maybe they "can't" hear it because they are focused on a different agenda. Maybe they are missing or misunderstanding a fundamental precept that's important. Maybe they have a different definition for one of the terms being used.

Most people, most of the time, want to understand. If communication isn't working, it's good to take a step back and ask what you might be able to do differently to help. It's not your fault, per se, but it is your responsibility, and it's good to come from a position of humility and act as though any disconnects are on you.

Is the team for you or are you for the team?
More thoughts on the inverted org chart

The inverted org chart is one of my favorite ways to think about leadership and humility. When we look at a "normal" org chart with the leader on top, we have a very strong, subconscious bias — that leader looks "elevated" and "better." We want to get to the top! Let's be ambitious! It's all about us, make the team "below" us bigger!

But the inverted org chart is a good way to remind yourself that the "leaf nodes" are really the most valuable and the management is, well, overhead. And it also gives us a different perspective on growth: not that a bigger team is "better" because you are closer to the top but that it's harder because you have to support more people!

Related to this, I've seen managers fall into a mental trap where they start to think of the team as existing to serve their needs. The main symptom of this is pushing

No Prize for Pessimism   175

someone into a role they aren't well suited or enthusiastic about because you have a gap and need it filled; it's all about your needs. But the team doesn't exist to solve things for you. As a manager, you exist to solve problems for them. The right approach is to put people not where you need them the most but where they will be the most impactful — which always means where their skills and enthusiasm are a good match.

The gaps that are left over are your problem to solve. Sometimes that means finding another person in the group who is a better fit. Sometimes it's hiring. Sometimes it's changing the problem spec or other behavior.

However you do it, just remember, the team doesn't exist for you or to solve your problems. It exists to be a team getting something done, and you exist to help the team with its problems. You are for the team, not the other way around.

Principles and discomfort

Almost every company or team has some kind of list of "principles." It seems like a reasonable thing to do — we should stake out a position that says who we are and what we value, right?

But these are usually more like goals or aspirations than truly helpful principles that can be used to make decisions daily. The reason is that they're mostly a grab bag of hard-to-argue-with ideals: quality matters, go fast, every voice matters, etc.

But the reality is, principles only really matter, and only really show up, when there is a conflict, a constraint, or a real choice to be made. Quality matters? Okay, but what happens when the schedule is behind, or there's an important demo or milestone or deal? Does quality go out the door then? If so, then the real principle is more like "quality matters unless we're in a hurry," which is less fun to put on a poster.

This isn't to say that teams shouldn't have cultural principles, or shouldn't write them down. But they should be written with an eye toward tradeoffs and priorities. One way to do this might be to list them in a strict order if you can. What's more important, speed or tech debt? What things do you explicitly contemplate not doing? Can you write those down? How does conflict work, both between team members or projects and between principles?

"Disagree and commit" is one of my favorite expressions of this kind of principle. If you unpack it a bit, what it's saying is: "We have a principle of listening and debating that's important, but being able to execute is even more important, so we agree that there is a point where, even if we disagree, we will do so explicitly and promptly, set it aside, and commit to moving forward together on what's been decided even if we don't agree with it." "Commit" is the principle that wins over "agree" or "reach consensus."

If you've built distributed systems, you've heard of the CAP theorem. It says that you can only have two out of three: consistency, availability (in a particular sense), and partition tolerance in a service. Sometimes you are in a state where that doesn't matter — the network is healthy, the load is good, everything works, and you don't have to give anything up right at the moment. But as soon as there is stress in the system — a server goes down, the network partitions, etc. — whatever choice you've implicitly or explicitly made in your design (essentially, the principles of the design) will manifest itself.

Similarly, if your company adopts principles that are fundamentally impossible to achieve all at once, you may have periods of low stress where they appear to all be followed. But there will inevitably be a period of higher stress and, again, those choices will manifest because they have to.

Successful organizations think carefully about these constraints and collisions, and design their principles not as a grab bag of good things that they hope for, but as an explicit, ordered (by importance), list of "must do" and "won't do" statements.

Chapter 8

Systems, Organizations, and Incentives

Look for the next bottleneck

Have you ever wondered why some things get faster, cheaper, and better over time while others seem to stay the same or even worsen? Why can we stream movies and video chat with people worldwide but still wait for hours at the doctor's office or pay a fortune for college tuition? The answer is optimization, or finding the best way to do something given some kind of limitation or constraint.

We often think of optimization as something that engineers and scientists do but, in reality, we all do it every day. We optimize our time, money, energy, and attention to achieve our goals and solve our problems. We look for shortcuts, hacks, and tips to make things easier, faster, and cheaper. We try to avoid waste, inefficiency, and frustration. We seek to maximize our value, satisfaction, and happiness.

But optimization is not a one-time thing. It is a continuous and dynamic process that responds to changing conditions and opportunities. Complex systems, like the ones we live in and interact with, will always have something that will be the next bottleneck or scale constraint. A bottleneck is a point or factor that limits the performance or capacity of a system. A scale constraint is a condition that makes it hard or impossible to increase the size or scope of a system. For example, in education, the bottleneck is teachers. Only so many teachers are available, and they can only teach so many students at a time. In medicine, the bottleneck is practitioners. Only so many doctors, nurses, and therapists are available, and they can only treat so many patients at a time. They are doing heroic work, and they need help.

Disruption is when a new technology, product, or service breaks or bypasses a bottleneck or scale constraint and creates a new way of doing things that is faster, cheaper, and better than the old way. For example, the emergence of the internet disrupted the bottleneck of distance. Before the internet, distance was a major constraint for communication, information, and commerce. It was slow, expensive, and unreliable to send messages, access data, or buy and sell goods and services across long distances. The internet removed or reduced this constraint and enabled new business ideas and tools, like email, social media, e-commerce, and mobile.

How can we look for and take advantage of moments when a bottleneck is disrupted or otherwise changed? One way is to ask ourselves: What is our system's next bottleneck or scale constraint? What is the thing that is holding us back or limiting our potential? What is the thing that is causing us the most pain or frustration? What is the thing that we wish we could do more, better, or faster, but we can't because of some limitation or obstacle? Then we can look for new technologies, products, or services that can help us overcome or bypass that bottleneck or scale constraint. We can look for new markets, customers, or users that can benefit from our solution or offer. We can look for new ways to combine, integrate, or leverage existing technologies, products, or services to create new value or efficiency.

As engineers and entrepreneurs, we should look for those scale bottlenecks and be sensitive to new tech that changes them. We should be curious, creative, and adaptive. We should be willing to experiment, learn, and iterate. We should be ready to embrace change, challenge, and opportunity. We should be optimistic, ambitious, and visionary. We should be optimizers, disruptors, and innovators.

Absolute vs. relative
How we measure and understand

As technologists and scientists, we are used to thinking about absolute measurements: Our room has a specific temperature, our bank account has a specific number of dollars, the bug database has a certain set of bugs, and so on.

But humans aren't that great at evaluating absolutes; we tend to think more in terms of relative measures. We don't think, "It's 65 degrees in here, I'm cold" (unless we look at the thermostat); we feel cold relative to a bunch of things—and a 65-degree room can feel really warm if you walk into it from outside in the winter.

Many of these relative measurements are essentially running averages. Wealth is a good example. If you ask most people how much wealth would make them content, the answer is usually roughly twice what they currently have, no matter what that amount is. We unconsciously keep a running tab, adjust to the usual state, and then compare from there.

This can happen in codebases and engineering teams. If you're not careful about measuring the quality of your code and product clearly and in absolute terms (bug count, latency, usage, etc.), it's very easy to acclimate and decide that something is more acceptable than it really is. The technical term for this is "because we always do it this way."

This is true for people, too — teams can slowly lower their standards in all kinds of ways without really seeing it day to day: civility, engineering standards, velocity. If you look around you and most people aren't working that hard, well, then in a relative sense, you don't seem so bad, right? It's very easy to build bad code or, worse, waste time in your career developing bad habits.

Not everything can be measured in absolute terms, and it's not always useful to do so anyway. But it's worth remembering: The frog that got boiled never noticed because he was running an average temperature calculation.

That can't be done, or we are already doing that
On receiving new ideas

Organizational cultures are infamously difficult to change. Jeff Bezos said that your culture is set after your first 100 hires because the mass of the existing culture is too much for any newcomer to change effectively after that. This has both good and bad aspects. Once you've found a successful culture, changing it can be detrimental. You have to be very certain that you're making positive changes. And, of course, the hardest kind of cultural change is when you've found a "local maxima," a ground truth in a team or group. Then you **have** to make negative changes to get to a new high. It's hard enough to get a large group of people moving in one direction when the goal is obviously better, but when it's a leap of faith that will be worse for a while — much harder!

One of the most classic examples of this kind of local maxima is the innovator's dilemma itself. Sometimes a company is being disrupted by changes in the market or technology around it. It has a successful business (the local maxima) that will be destroyed by this change over time. So they need to change! But any move to the new world of the future reduces revenues today, which is painful! So they're stuck – until some huge pressure or culture change allows them to follow a path to the new configuration.

It's also very natural for organizations to be filled with people who have "selected into" the culture. That's the point of the Bezos idea above: Once the group gets big enough, it's very hard to be in dissonance with the overall culture. So there tends to be a lot of reinforcement.

In that context, challenging ideas are received in one of two ways: rejection for some fundamental reasons that seem inarguable (i.e., that can't be done) or

186   Sam Schillace

rewritten into the "rules" of the culture and neutralized (i.e., we are already doing that, so nothing needs to change). To some degree, these two poles are always "right": There are always challenges and problems with any new idea (so it's easy to say "that can't be done") and, in a large enough organization, there are always projects or at least large groups where any new thing might fit, or have principles that cover it, particularly if the manager has done their job well (so "we are already doing that" would be true).

It's challenging to react to new ideas in a different way, but it's helpful to be aware of the pressures toward the status quo. One way to break out of this mentality is the "yes and" mindset. This is something used in brainstorming and improv. The only rule is you can't reject the idea; you have to say yes "and" — add something to it. So, for something that seems uncomfortable and impossible, we try to look at ways it might be approachable or things that may have changed that make it more achievable. For the other side of the coin, we have to look honestly at how the new ideas would fit into the existing work and budget; it's not enough to say we "could" do something; we have to be able to firmly say we are doing it, and that involves deeper thinking about what does and doesn't match.

This is all challenging. Large groups always have full budgets and well-defined agendas. New ideas are mostly a distraction and usually aren't useful. But it's not healthy for any culture to stop listening and incorporating new ideas, and if it's very challenging to change, it's worth thinking about whether the culture has found a local maximum that isn't really optimal.

Goal vs. process

Imagine if you had to drive your car to the store by giving it a complete set of instructions upfront, with as many contingencies as possible. So: turn the wheel 13 degrees, accelerator to 10% for 15 seconds, brake until 15mph, etc. I wouldn't want to ride in it! The odds of the car getting to the store are pretty low and only non-zero if the environment is simple and has few surprises. This is waterfall development, and process orientation.

Instead, we iterate our way through the process interactively, making many adjustments along the way. We have a capable agent in the car (us) that can interpret many changes and react to them in real time, and we are (mostly) safe getting to the store. There are still accidents and mistakes, but the iterative, goal-oriented approach is much more robust.

As it is for cars, it is for software teams and even software products. For teams, the analogy is obvious, and we tend to do better if we are goal-oriented instead of process-oriented. "What problem are we solving for the user or customer, and when?" rather than "Here's a spec, implement it." Goals have value all the way through the team. To the degree that the team understands the goal and intent of the business and product, it can act much more effectively.

And it's true for products as well; to the degree that we can help users with a goal rather than presenting them with a rigid process, they will be happier and more effective. This means designing with intent in mind, and with a tool and modular mentality: Give the user well-defined behaviors and pieces that compose and combine. The best feeling is when a user surprises you with something you didn't think your product could be used for.

Users really only care about goals. To the extent that they care about process, it's to have as little of it as possible to get to their goal. Teams, too. The best thing you can do is be as clear about goals as you can be and to help your team and users find their way effectively.

Context matters
That which is measured is built

Building software (or anything complex) is challenging for several reasons. One of the central themes, though, is the tension between the short-term and long-term perspectives. This manifests in many different ways, often in design; for example, the quick hack gets you going but, in the long term, becomes the dreaded "technical debt."

But debt is kind of the wrong metaphor here; you didn't "borrow" some tech that you have to pay back. It's more that, in the early phases, you focused on a narrower set of problems because you needed to. The error wasn't the focus per se; it was that the narrow context was focused on for too long.

But starting out with a very broad focus ("long-term architecture") can also be lethal. There are many teams who have failed to succeed because they didn't spend enough time on the more immediate context of product and market fit, and wound up either being late or never finishing.

Context also matters when looking at metrics. It's always easy to find some kind of vanity metric that will tell you the story you want to hear — just like it's easy to feel good at the end of a day of hard coding on a "hack" that gets the wrong thing done. It's really critical when using data and building metrics to find the metrics that really reflect your business goals. Maybe feature count is really important to your sales team, but performance and stability are important to your users. Watching engagement might tell a better story than just watching sales if you're not watching for churn. It's easy to get lost in any of these numbers — you can watch latency and feel great about product health, but if no one is using the product, it doesn't matter.

Context is hard to get right but is really critical in a number of areas of software design. We can only focus on one context at a time, so having the meta-cognitive ability to think about which context to focus on and to actively switch among them to get the right mix is a critical skill for engineers and business folk alike.

Fear and metrics

Let's revisit the tension between short- and long-term thinking. Navigating this balance successfully is a critical part of being a good engineer or entrepreneur. Some of the challenges are fairly obvious: investing in code design versus quick hacking, for example, or generalizing patterns versus copy and paste.

Some of them are more subtle, though. One of these has to do with how teams engage with product metrics.

There are lots of different ways to measure the success of a product: latency, engagement, growth, stability, etc. Let's pick growth to focus on.

There are a few ways to measure growth. One is absolute: How many more daily active users do we have this week, relative to last week; that gives us a percentage growth rate. The other way is to look at share: What's the market we are in, and how is our share of active users changing in that market?

The first one is easier to "lie" about; for example, during covid, lots of products increased absolute use but decreased relative share. Here's the short/long tradeoff though: It's easier and more comfortable to pick the version of a metric (growth in this example, but any metric has different forms like this) that tells you the story you're comfortable hearing.

But that comfort is the short-term thinking. It might make it less disruptive to your group or company in the short term, but it's not really healthy in the long term.

As is usually the case, there's no simple answer to this problem, though there are tools in the toolkit: setting up pre-commits (in this case, measuring the long-term objectives and key results that really matter to the company), building complementary processes where some external factor (an analytics group, say) is explicitly tasked with making sure the hard questions are being asked, committing in public, etc.

It's subtle, but it's important to look for ways in which we trade off near-term comfort for long-term health and growth.

Let's cut some cake
How to make decisions algorithmically and stop fighting

Quite a while ago, someone came up with a clever, mathematically provable optimal solution to the age-old problem of how to cut cake. This might not sound very useful, but it is, and it applies to engineering, too.

The problem is this: Suppose you and one other person want to share something (like a cake, for example). How do you divide it fairly? You can't necessarily trust the other person to decide, and they can't trust you. No one can be trusted to decide! If there is no other authority, what do you do? You have to find a way to align your incentives somehow — and that alignment problem is the thing that's more generally useful. You can decide "together" if you break the decision up.

The (pretty obvious) solution is this: One person cuts the cake into two pieces, but the other person gets to pick who gets which piece. This motivates the first person to do the cutting as fairly as possible, which then aligns the incentives: Both people want the cutter to be as fair as possible, no matter who the cutter is!

In engineering teams, there are often similar problems. Not resource contention, per se, but arguments about what the right approach to a technical problem is. In some sense, it's the same problem; two balanced groups are contending for the same resource (the time of the team). So step one is to try to reframe a problem like this in terms of a "dividable" resource.

There isn't as clean an algorithm here, but we can look to cake cutting for inspiration. Maybe one person splits tasks into groups; the other decides which approach matches which tasks, and then you conduct an experiment to see which approach works better. Or maybe one person breaks the **team** into two halves, and the other person decides which half gets to **make the call** on the approach (recurse until you get to a three-person team, and one person winds up getting to pick for everyone). That incents the first person to make the deciding teams as fair and representative as possible.

This algorithm also generalizes to more than two (let's call it n) people. In that case, the first cutter cuts the cake into a 1/nth piece and a (n-1)/n piece, and the second person decides to either take the small piece or be the cutter for the big piece (and give the small piece to the first person). That's an example of using recursion when the problem is more complex.

Keep this technique the next time you get in a head-to-head argument with someone. Is there a way to "divide" the decision that lets you be aligned? Sometimes a little math is better than a lot of arguing.

Desire isn't a strategy

Most of what is presented as "strategy" in most organizations isn't a strategy. It's usually something more like a desire. Sometimes they can be a little mixed together — "build software users love" is *kind of* a strategy; at least it tells you something not to do (build crappy software), but it doesn't really tell you much about how to do it or why.

"Get rich" or "Get promoted and be successful" are the kinds of things you hear people say when you ask them about their goals and plans. That's fine, as far as it goes, but it's important to realize that those are relatively undifferentiated goals. Something that starts to feel more like a strategy is "specialize in field X, so I advance quickly" or "learn about startups and build skills to build something successfully."

Some big companies have obvious strategies. Apple has a strategy of "the hardware and software each differentiates the other," and this motivates the tactic of both investing in hardware (M1 chips, etc.) as well as user experiences (design). Amazon has a strategy of "take advantage of scale," which motivated the creation of AWS. Each of those companies also has *goals*, but those goals are less differentiated: "Get big, make lots of money, be successful in market X."

In larger organizations, it's very easy to get into the mindset of "We have to do something; this is something, so let's do it." Established organizations and companies have their own momentum, so it's easy to do something that is available and think that it's going to have some impact when, in fact, it doesn't serve a clear strategy. This often shows up as "strategies" that aren't actionable and are really just desires. Good strategies help you make choices. If your strategy doesn't tell you what NOT to do (it just says "do all of the good things") or it doesn't help you form tactics to get things done, it's not really a strategy; it's more of an aspiration or goal.

Manager or short-order cook?
Does your organization really value quality?

It's straightforward for organizations to have "values" that are uncontroversial, things like "user focus," "quality first," etc. Without real commitment, these are just nice things to say, though. We've all been in organizations that say they have a certain kind of value but don't really.

One way to understand if your organization really values something is to see whether it will make tradeoffs to get it. Quality is a good example. Whether engineers and engineering managers are empowered to ship, or not ship, based on quality concerns tells a lot about the real priority. And here, quality is a broad term that encompasses things like bugs but also just overall user experience.

One low-quality pattern is the "short-order cook." In that model, engineers and first-level managers are given a series of tasks to complete with very little context. They are measured almost entirely on throughput; quality only enters the picture insomuch as a task is returned to be worked on again, and it lowers overall throughput accordingly. These organizations can be effective in execution but rarely build strong user experiences; engineers and managers usually best understand how well something is working, and removing their ability to object to a bad product (and introduce a delay) creates all kinds of problems.

This is an interesting question to ask when you're interviewing, and it's an interesting pattern to look for

No Prize for Pessimism   197

if you're a leader. Can engineers hold up a release for quality issues? Usually, the answer is yes for the most egregious examples, so then the real question is: What severity of issue is the smallest that an engineer could delay a release over? This is now a kind of uncomfortable question — and now you're learning something valuable by answering it!

Watch for the short-order cook syndrome — both as an engineer and as a manager.

Hero culture means your systems are broken

We like to celebrate heroes in software. There are always those engineers who save the day: the only person who knows how a build system works or the one who grinds long hours in response to a security panic, fixing all the bugs by hand. We like to celebrate them. How many times have we heard an executive praise that kind of "hero" behavior?

But it's actually a really bad sign if your organization has heroes. Chef Tomas Keller was once asked why his kitchen looked so calm, and he said "because we are prepared. If you have to rush, you're already lost." It's true here as well. Hero culture means something is amiss with your systems or incentives. You wouldn't need the hero to rescue you if you had built a healthy system with good infrastructure, readable documentation, repeatable processes, and so on.

This isn't just true for individuals, though usually they're the ones most motivated. Sometimes teams or divisions of companies can act in this role, and that can point to similar issues with business models, culture, and more. I heard a story once about a team at Apple whose feature was cut from a release because it missed the deadline, so they worked all weekend to get it done. The manager coming in the next morning was annoyed: They broke the process; the decision had been made. That feature didn't make it into the release, and that's a great example of being disciplined about system and process and not letting hero culture take hold.

It's natural and nice to be grateful when someone saves us, and we should be. But we need to separate that natural feeling from the understanding that we shouldn't have needed that effort. The right way to respond to a hero is "Thanks, that was really helpful. Now let's all work together to make sure it never happens again."

Flexibility vs. randomness
'Ah, ADHD, my old friend'

A challenge of any creative endeavor is balancing being stubborn about your vision and being flexible about learning. Neither extreme works well, especially in entrepreneurship, where it's critical to respond to the market but also critical to not be so responsive that you lose your insight and become undifferentiated. This is true for most creative fields, too.

I have ADD (ADHD but without the hyperactive part; I can sit still for the most part, but my brain hops around a lot). It's a real gift when it comes to being creative and doing lateral thinking, but it can also lead to some real randomness. It's hard to know the difference sometimes. Good ideas are usually initially unintuitive based on what you're doing at the moment or a connection that seems unlikely but pans out. Sometimes, when you are stuck on something, coming at it from a different perspective helps. But sometimes, it's just procrastination.

It's hard to grapple with this problem. Again, neither extreme is really optimal; the right mix is somewhere in the hard middle. So, you have to attack it less with a formal rule and more with heuristics, like "Do the important and urgent stuff first, then important and not urgent," etc.

Sometimes, I find that timeboxing or budgeting helps. Like, I get to say or do three random things per day or per meeting, and I try to keep myself in that box if I can. Another thing is to try to stop and ask, "Can I plausibly point out how this advances me toward the current goal?" If not, then at least give a limit to the amount of time spent in that state. It's okay to do something that "feels right" for a while, even if you can't explain why, but too much of that begins to tip over into randomness.

One other strategy I use to manage this, related to budgeting, is "parking lot": I'll let myself have random ideas but simply record them somewhere without doing much and come back to evaluate the list when I have time. Often, I find that time makes things more precise in both directions. This lets me have all of the creative ideas I want without being disruptive. This works well in meetings; I take notes and often just don't say what is in them because they become less relevant by the end.

As with many things in our professional lives, there is a balance here. With AI, where there are lots of new tools and techniques to use, it's easy to become lost either by being so rigid in our workflows that we miss out on something new and powerful, or by being so enamored of chasing all the new ideas that we never get anything done. Striking this balance is hard, but critical.

Chapter 9

Career Advice

Do the more rare thing
Usually, at least

We often face complicated decisions with lots of factors to them. We all struggle with these. Sometimes, you can line up the factors and make a rubric to help you decide. Picking a college? Well, what matters to you? Weather, academics, location, surrounding city? You can rank each of these, say, from 1 to 10, and add them up. It's not perfect but that gives you some kind of idea of which one you like better.

That's a good example of reducing the dimensions of a problem so that you can impose a strict ordering (metric) on it, but often that reduction doesn't really capture the whole story. They might tie, or there might be a factor on each side that's really strong, but it's different in each case, and you're back to either/or that's fairly balanced.

However you get there, if you have that kind of otherwise balanced choice, one nice rule of thumb for breaking the tie is simply to take the option that is rarer (less common or harder to get). The idea is fairly simple: If the other choice is less rare or easier to get, you're more likely to be able to reverse your decision in that direction. Also, you may not be able to get the rare option later. Part of making a choice is being worried about regretting the choice, and this is a small amount of insurance against that.

This isn't universal or foolproof by any means — sometimes it's also hard to tell what's "rare"! But it can help in surprisingly many circumstances. It's a nice external forcing function that gives you a way to choose without having to agonize about it too much.

Work isn't what you think it is
Some unexpected career advice

If you get to a particular stage of your career and you're lucky, young people will start asking this simple but devastating question: How did you do it? (And, really, they're asking how they can do it). If you're honest with yourself and you try to give a really useful answer, you'll quickly come to realize that your career, like everyone's, is full of random changes, non-obvious choices that turned out to be important, and just plain luck. So it doesn't generalize — no one can follow someone else's path.

You can't step in the same river twice, but we navigate rivers, right? So, there has to be some answer to this puzzle. That answer is **impact.** Having a lot of impact in whatever you are doing, however you define impact, is usually what leads to long-term success. All the better if that impact compounds somehow: building a business, or a reputation, or a practice that feeds on itself.

Okay, but how do we get to this idea of compounded impact? The impact can mean many things; it's different for a programmer, a designer, or an artist. But it always comes from the same place: passion. Having a passion for what you do so that you're obsessed with it, constantly want to get better at it, and think about it all the time is how you develop impact. It's really important early on when you aren't very skilled because passion keeps you coming back to it. And it's important for the compounding effect. Passion keeps you investing and building.

It also leads to something really counterintuitive sometimes. We have a mistaken tendency to equate suffering with adding value. It can't be work if it's fun, right? It must be a struggle to "count" as work. But this is patently false — many of the highest performers have loved their work, and it's the best place to be, mentally.

But it's also kind of slippery — if you're really good at something, and you have passion for it, it's very easy to feel guilty that you are getting paid to do it. This connects to impostor syndrome, too (more on that in the next couple of letters): It's really easy to feel like, "Gee, I'm off track because work is fun and doesn't seem hard, so it must not be valuable." The right word for that, to quote Charlie Sheen, is "winning." If you are good at something, it will often feel easy to you. That's a great sign to look for that you're actually on the **right** track — that's something that's a fit, that you have skills for, and that you are passionate about.

That will lead to an impact and a satisfying career.

Proof is in the doing
It's easy to fool yourself

Most people fight against impostor syndrome — the unshakable feeling that you don't know what you're doing, that you'll be caught out. And it's often the case that the most capable people feel it the most, so most of us consciously fight against it and try to build confidence where we can.

But it's easy to go too far and confuse familiarity with capability. We spend time around lots of complex subjects, and we work hard to stay informed and current.

We learn the vocabulary of our teams and companies, understand the outlines of the technology, sit in meetings, and hear decisions being made.

It's easy to confuse familiarity with competence. When I was studying math, I saw this all the time; it was easy to look at a math problem and think, "I can solve that." About half the time, that was correct — usually, it was harder than it seemed. Proof is the only ... proof.

Another way to think about this is languages. You might be around people who speak a language that you don't speak. You might understand a bunch of things about that language: how it sounds, how it looks when written, even some of the words. But you would fail if you actually tried to speak it. It's easy to do this with technological subjects, too — things like ML and crypto are easy to understand in the abstract without really getting into detail in a way that is usable.

It's okay to just be familiar with complex subjects. Often, cursory familiarity is useful in making decisions and understanding choices. In fact, this is a requirement for becoming more experienced in tech; there's just too much to know everything in detail, but you still need to understand the implications of things you aren't working on directly. So this peril is ever present — optimize your attention (always), but be careful not to fool yourself into thinking that you understand things better than you really do.

Self-fulfilling worry

Have you ever felt like you don't belong in a particular situation? Like you're not qualified or experienced enough? As if you're a fraud who will be exposed any minute? If so, you're not alone. Many people suffer from impostor syndrome, the feeling of being inadequate or undeserving of their achievements or roles. Impostor syndrome can affect anyone, regardless of their age, gender, education, or profession. And it can have negative consequences, such as lower self-confidence, higher stress, and reduced performance.

But what if impostor syndrome is not only a result of our self-doubt but also a cause? What if by worrying about how well we're doing, we're actually making ourselves do worse? That's the paradox of impostor syndrome — it can become a self-fulfilling prophecy. By focusing on our flaws and failures, we're reinforcing our negative beliefs and limiting our potential. We're also wasting energy that could be better spent on improving our skills, learning from our mistakes, and celebrating our successes.

That's why it's better to just focus on doing our best rather than comparing ourselves to others or setting unrealistic expectations. Doing our best means acknowledging our strengths and weaknesses, setting realistic and specific goals, seeking feedback and support, and embracing challenges and opportunities. Doing our best also means being kind and compassionate to ourselves, recognizing our achievements and efforts, and accepting our imperfections and limitations. Doing our best does not mean being perfect, or being better than everyone else, or never making mistakes. It means being authentic, resilient, and curious.

By doing our best, we're not only overcoming impostor syndrome but also applying a system thinking approach to our personal and professional growth. System thinking is the ability to see the interconnections and patterns among the elements of a complex situation and to understand how they influence each other and the outcomes. By doing our best, we're not only looking at our individual performance but also at how it contributes to the larger system we're part of, such as our team, our organization, our community, or our society. By doing our best, we're also creating positive feedback loops that reinforce our confidence, motivation, and learning. By doing our best, we're also adapting to changing circumstances and embracing uncertainty and diversity.

Impostor syndrome is a common and understandable phenomenon, but it doesn't have to hold us back or define us. By shifting our mindset from worrying about how well we're doing to focusing on doing our best, we can overcome impostor syndrome and enhance our system thinking skills. Doing our best is good not only for ourselves but also for the systems we're part of. Doing our best is not only a personal choice but also a social responsibility.

Stories we tell ourselves

This is a little bit more of a personal letter. I was watching a video of someone with Tourette's, and they talked about how important the diagnosis was: Until they understood what was going on, they self-labeled as a bad person, someone who was out of control, did bad things "for no reason," and was "broken." Now, they still have the condition, but they understand it, are open about it, and have worked it into their life very successfully. All that changed was the story they are telling themselves.

I had a similar but less extreme experience with ADHD. I wasn't diagnosed until my 50s. Until then, the story I told myself was that I was just weak-minded, scattered, unfocused, and emotional. Understanding

ADHD means I know where all of that comes from. Sometimes it lets me recognize that a sudden emotion isn't really valid, and lets me step aside from it. But, more importantly, it has let me really understand, and own, the fact that my wonderful ADHD brain gives me all kinds of superpowers that other people don't have: creativity, an incredible facility with visual thinking and abstraction, speed and insight, and so on. I don't feel the shame I used to feel — the story I tell myself has changed.

Those are extreme examples, but we all tell ourselves stories about what we can and can't do. "I'm not good at math/art/music/people" or "I'm impatient" or "I'm lazy." There are two things to think about here. The first is that you should try to look for root causes and systemic effects rather than just symptoms. If there really is a behavior that's fundamentally different, try to understand it in a neutral way if you can. Dig into what the causes might be.

The other is a bit more zen, like the parable of the farmer's son. If you think something about yourself is bad just because it's different, how do you know? Can you tell a different story? ("I'm not good at infrastructure" can become "I'm a great front-end coder," for example). Is there a place where your natural skills and tendencies are valuable instead of a liability? Much of life is noticing

No Prize for Pessimism   211

this about ourselves and adjusting to find environments where that is more true.

Stories are most of what makes us human. We tell them to each other all the time, and we tell them to ourselves most of all. Theoretical physicist Richard Feynman said something like, "You must not fool yourself and you are the easiest person for you to fool." Try to watch yourself today and see what stories you are telling yourself without thinking about it. Once you've spotted them, think about how they might be reframed to be more helpful.

Planning for 35 years

I spent some time last week (October 2022) at my alma mater, the University of Michigan. It was a very humbling experience — the school uses Google Workspace (which I still call GDocs), and the students were amazed to meet someone who had a part in creating it. If my overwhelmed freshman self knew back then (over 35 years ago) that I'd someday come back and the whole campus would be using my software, I think I would have hidden under my bed for a week!

There's a question the students asked in almost every setting: How did you plan your career? Usually, when I am asked this question, I say that I didn't, that it would have been impossible to predict all of the changes 35 years ago.

In this setting, it was even easier to say that because most of the things I'm working on now, like AI, were around in only rudimentary forms, if at all, back then. The first three startups I did were pre-internet (it existed, but it had maybe 100 people on it in academic institutions). The iPhone in my pocket is roughly the equivalent of 25 to 60 Cray-1 computers (depending on how you count), which was the unreachable cool computer when I was just out of school.

It's a lot of change. The past 35 years were chaotic and hard to predict, and I think the next 35 will be more so. Many exciting changes are happening technically right now, and we probably can't see most of the really interesting ones that change now will unlock in 10 or 20 years.

The advice I gave is more process than goal. You want to learn to learn because it's true that you won't have all the skills now you need in even 10 years. I remember fussing about whether to learn C or Pascal. Ultimately, I wound up learning to code in 10 languages for money (and more for hobbies), and it wasn't an important aspect. Pay attention to the things that feel easy to you, where you don't know why people want to pay you for something that might even seem fun; we tend to undervalue things we are good at because they are easy to do.

Keep learning. Find the place where you add the most value because of your interest or your natural skills and invest in a way that compounds. (In other words, don't waste time on things like social media that don't add up; learn a skill and then stack it on top of another skill, and another.) You'll make good career decisions and bad ones, but if you build a mechanism to be aware, be adaptive, and compound value in the things you're good at, you'll do fine, no matter how choppy the sea.

New tools or old?
Don't be too comfortable with your tools

When you're new to engineering, there's a strong (and healthy) desire to learn as many new techniques and technologies as possible. This is great! It's good to have lots of tools in the toolbox and to know what's out there, and we should always be exploring.

But there's a counterpressure here too, which is that it's often more efficient to work with code or tools that we know and understand well. And this also makes a ton of sense: Why learn everything from scratch each time? But then we risk becoming stagnant or isolated over time, as the rest of the industry moves on.

So, how do you balance these? One part of the answer is that you have to consciously do a little of both, all the time. Have tools and code you invest in and are comfortable with, but don't let them be the only things you use. Working with something new is almost always less efficient, so you have to make space in your life for that inefficiency. You can call it "play" or "research" or "investment" if you want. But some regular percentage of your time should be dedicated to trying out new tools and sometimes incorporating them.

Another part of the answer is to look for tools that compound over time: they compose with each other, the capabilities are reusable and can be applied elsewhere, etc. Being very familiar with a scripting language or development environment might be a good example of this. Sometimes engineers build their own toolkits of scripts and other things that are useful — this is a good way to get compounding, if the tools are general purpose enough to be reused.

All the tools in your mental toolbox should be there for a reason. They should be as useful as possible: You should have a great understanding of them, use them regularly, and get compounded value from them, but you shouldn't be trapped by them. And you should discard them when something better comes along.

No Prize for Pessimism   215

It's essential to be comfortable with our tools, but not **too** comfortable.

Managing ... up?

Someone asked me on LinkedIn if I would write a letter about "how to manage up," so here are my thoughts on that.

One way to approach this topic is to think about politics: how to figure out what your manager wants and tell them what they want to hear. This is a reasonable strategy for corporate survival and promotion, I suppose, but it doesn't really get you very far in terms of trust and building effective teams. And good managers are very good at sussing out when someone is doing this, so even though it might seem like it helps you advance in your career, it can label you as someone who doesn't have their own opinions or add much value beyond syco-phancy. Good managers want to know what you think, not what they think.

There's another way to think about this problem, though, which is more constructive. In general, you should always think about and try to solve the prob-lems at the level above you. This not only helps the entire team be more effective, but it also helps you grow in your career: Trying to understand how your boss views

216   Sam Schillace

the situation is good training for moving into that role eventually. And good managers recognize when this is happening and usually encourage it.

Finally, if that's done well enough, you can start to anticipate needs and structure your work product accordingly. For example, most folks early in their careers will produce very long, detailed reports or emails in response to a question — usually with the idea that they are showing off how competent and smart they are. But the reality is that your manager (like everyone else) is busy, so taking the extra time and effort to distill something down into a much more easily consumable piece is really valuable to them. The best kind of direct report is the one you can trust to filter out all the details that don't matter and present to you just the parts you need to understand, why you need to understand them, and what they think the right thing to do is. It's hard to do well, and it does take longer, but again, it's practice for being at that next level yourself.

Manage up not by playing politics and saying "yes" but by working to understand the perspective of your manager and helping with their workflow and problems.

Asking vs. telling

One of the biggest challenges when working with legacy technology or established teams is challenging the deeply held beliefs in the code or the people (usually both). The naïve thing to do (which many of us do early and even, embarrassingly, later in our careers) is to tackle it head on. We see something we think is wrong, and we firmly blurt out the "right" answer. Sometimes, we even dismiss the other party as ignorant, stubborn, or otherwise unhelpful. Usually, the result is predictable: The other party digs in, fights back, and doesn't listen; everyone gets defensive when challenged directly. If it's code, you may easily find yourself lost in edge cases and complications that arise from the assumptions you are challenging being reflected in the rest of the system. All of this makes it harder to get change to happen.

The answer, of course, is listening. But that's not really super helpful; if you just passively listen or passively read the code, it's not really going to change anything, right?

Instead, there's a form of inquiry and listening that's more active and can be used to advocate in some ways. It's usually referred to as "Socratic questioning," but it goes under other names like "the five whys." This involves asking questions that really take the other perspective into account and are honestly curious.

The real trick here isn't to ask questions that are obviously biased, aggressive, leading, or otherwise hostile (the "When did you stop beating your dog?" kind of question). The real trick to this is to cultivate genuine curiosity and empathy in those questions, which means you have to really **be** listening and be willing to change your mind if you learn something new.

Usually, when something seems wrong to you, you'll have mentally skipped ahead to an outcome that you think will happen based on the decision you think is wrong. Try to ask genuinely curious questions that lead to that outcome rather than simply jumping to arguing about it. Think a particular approach won't scale? It's not useful to just state that, but it might be useful to ask questions like "How does this system work when there are 100 million users? What's the bottleneck? Why do you think this is the most important design principle?" etc.

This approach works with code too: There was intent, and dogma (even if it's just the dogma of laziness) when code was written. Interrogating it honestly can help you understand what problems the coder was trying to solve and what is feasible to change.

None of this is perfect, of course; people can be stubborn, and code can be irredeemably broken. But conversations that start with a genuine desire to listen and learn and have some genuine empathy are much more likely to make progress. Very few people do things that they consciously think don't make sense; everyone has a context and goals, implicit or explicit, that constrain and inform their behavior and beliefs, and everyone is solving that set of constraints in a way that makes sense to them. Trying to invalidate that context only makes them angry and insecure. You have to understand the totality of the reasoning that got them to their position, not just the end position itself. Listening, understanding, and trying to solve problems within that context at least gives you a chance to change your mind.

And I think this is true beyond just code. Humility is an underrated quality. We think of scientists and leaders as heroes (which they are) but we seldom think of them as humble. Humility is the quality that lets you "ask instead of tell" and often leads to insights and trust. Good scientists notice when something doesn't fit their theory and humbly ask why. Good leaders hear when someone they lead has a good idea that doesn't fit their theory and humbly ask to understand.

The power of 'Yes!'

Early in my career, I was the typical headstrong young engineer: I knew what I thought was right and I wasn't shy about it. As you can imagine, there was a lot of "no" in my interactions with other people. There was not a lot of listening or collaboration. Since I was stubborn, quick on my feet, and smart, I tended to feel like this was successful and, to some degree, it was. But it's a very limited approach to getting things done. Let's explore a few examples.

At one point at Google, another team wanted something from my team that I didn't really want to do. We argued a bunch over it and weren't really getting anywhere. Finally, I reversed course and said "yes." Still, I added some of my concerns into the yes — we had to broaden the scope so that the solution was more general. I wanted obligations from the other team in terms of support, and I had a bunch of questions that they needed to work to answer. The other team decided that the approach they were pushing wasn't actually what they wanted, and we found a different path. None of what I had asked for was done maliciously. They were the real concerns in my head then; I just hadn't been saying them, using "no" as shorthand. But saying "yes and" instead of just "no" let us have a real conversation.

Here's another example, something I'm still embarrassed about. I took a negotiating class once. There was an exercise before lunch where you role played either being a musician's manager or a radio executive, trying to cut a deal. I happened to get someone I knew well as my partner, and we quickly found some easy middle ground. I felt happy, and we went to lunch early.

But the real point of the exercise wasn't just to settle on that middle ground but to think about how else each party could help the other. The other groups that spent time on the negotiation found much more value together than my friend and I had. Sometimes, a huge amount more. I had thought we were in an easy "yes" mode, but in fact, we were saying "no" to the harder conversation, and we completely missed the point. Saying "yes" and really engaging in the negotiation together would have given both of us much more than we wound up with.

Sometimes it's subtle to understand if you're saying "no" or "yes and." Yes is almost always a better place to start from. It goes by many other names, like "growth mindset," "inclusiveness/empathy," and sometimes, just plain "listening."

Yes is much more powerful than no. Yes leads to your own growth as well as others, while no keeps things static. Yes is a path to conversation and understanding; no is a closed door. Remember, everything we have, everything we do, was once new. Someone somewhere said yes to it first.

Which derivative are you?

I like math! (sorry if you don't). Sometimes, it's easier to think about things using good analogies and tools.

Derivatives are one of those tools. They're a way to think about rates of change of functions — the "0th" derivative is the function itself, like a line or a parabola, etc. The first derivative is the rate at which the function changes at any given point. For a line, this is easy: It always changes at a constant rate. For other functions, the change itself can change: It can accelerate or decelerate. The "rate of change of change" is the second derivative. In physical terms, the function might be "location," the first derivative (the rate at which the location changes) is "velocity," and the second derivative (the rate at which velocity changes) is "acceleration."

These are useful things to think about in terms of technical organizations and efforts, too. Conversations are often focused on what amounts to "0th" derivative issues; for example, the current state of technical debt, or developer velocity, or bug count.

But the interesting conversations are all to be had at least one order up, and maybe two. Teams will often say, "Well, the bug database is impossibly big; we can't do anything about it." But a better conversation to have is, "Well, let's make it so that at least it's getting smaller, even if slowly (first derivative)." And then you can start talking about and managing how quickly it's reducing, which is the second derivative. These are increasingly mature and sophisticated ways to manage an effort — and usually with increasing returns if done well. Teams that not only manage developer productivity but think in terms of accelerating it over time, and even accelerating that acceleration, get very productive.

Much of software engineering is balancing long- and short-term needs to achieve overall productivity over time. Long and short term are different ways of talking about derivatives. The short term is really about position and the long term is about velocity and, ultimately, the change to that velocity over time (which we want to optimize). Higher-level derivatives can make some complex problems easier to understand and debate because they are usually more abstract.

Watch your thinking and try to ask: Which order derivative am I thinking about here? How can I move up the stack and think more broadly about rates of change?

Why do you want to convince them?

It's tough to change someone's mind. We all know this, and we've all had the experience of getting stuck with someone who's really defensive and refuses to listen to our increasingly impassioned arguments (or we've been on the receiving end of this where we really dig in hard and don't want to listen to the other person).

This is a well-known phenomenon that happens for a bunch of reasons, but one of them is that people are skeptical of each other's agendas. And, in particular, if you don't know what someone's agenda is, you will make one up. We're all a bit paranoid, so we will usually find a "worst case." And that leads to defensiveness.

One way to get past this is to ask, "Why do I want to change this person's mind?" and answer it for yourself as honestly as possible before engaging — and then be upfront about it! In technical arguments, there's often an aspect of politics or status or ego involved; the arguments are less about the tech and more about the value or status of the people involved.

So if you can ask yourself this question, and the answer is something like "I'm worried your design will be too costly in resources" or "I'm worried I know about some issues that I don't think you're aware of that will impact the decision," you can often find your way to a more collaborative and helpful posture. Someone will react much better to "I am worried you don't know something and will waste time on this design" as opposed to them inferring an agenda like, "I think you're dumb, and

I'm trying to expose you and hurt you" (which people really do unconsciously think!).

If you can genuinely understand your motivations, and they're genuinely meant to be helpful to the other person, expressing them is a great way to take some of the defensiveness out of the situation. Now you're both on the "same side" looking at the challenge together.

How to work with the best people
"Users are lazy" in another realm

When I was speaking at the University of Michigan, students asked me about various aspects of this ongoing letter-writing practice of mine. One thing that came up a few times was the idea of trying to work with the best people you can. I was asked how to make that happen.

First of all, why should you make it happen at all? Well, we often "don't know what we don't know." This manifests in many ways: sometimes in plain ignorance and sometimes in the more confident form that gets labeled as the Dunning-Kruger effect, where the less competent people are actually more confident because they aren't aware they lack the skills to evaluate them-selves accurately. Being around people who are higher on the skill and performance spectrum can give you that context, which is needed to improve.

So how do you do this? Good people also want to work with the best people possible, so you're pulling in the wrong direction. This happens in music jams, too: Everyone wants to play with the best jam, but those people also want to have a great time, and they don't want someone "busting the jam."

This is a variant of "users are lazy," I think. Going to someone who has a better skillset and expecting them to let you work with them without getting anything in return is akin to building a not-very-useful app and expecting users to make some effort to use it, just because ... I don't know ... you're a nice person?

In apps, you have to find a way to come to the user and deliver value to them without them having to exert much effort. In a jam, you have to bring something good musically, and you have to show upfront that you know how to add to, not subtract from, the experience. In apprenticeships, you have to think hard about what the person you want to work for needs and try to show up with that value obviously in hand. It can be small things; usually, it's tasks that are less effective for the person to do themselves, but it can be other things like energy, new ideas (careful not to be overly confident or forward at first!), or even just that they want to give back and you present as someone receptive and easy to work with.

No Prize for Pessimism   227

It's always worth spending time understanding what the other party needs, and then putting it into terms they can easily consume — whether that's a break in a bluegrass jam, a bit of code in an OSS project, or an article on the internet.

Work with the best people you can

Younger engineers often ask me how to set themselves up for success in the tech industry. It's always a moving target. Folks are usually asking something like "What skills should I have?" or "Should I learn this language, or be this kind of expert, or do frontend/backend/whole stack?"

There are lots of things you can do to raise the odds of being successful. Practice your craft (coding, design, writing, whatever) as much as you can. Build things, challenge yourself, and have a growth mindset — these are all pieces of advice.

But the one I like the most is something I learned from playing music. When you play with other folks, the goal is always to find a group of people who are all better than you; the ideal group is the one that just barely tolerates you. If they were any better, you couldn't learn from them because they wouldn't want you around, or you couldn't hang with them at speed. You want really good musicians you can learn from, but not so good that you can't.

But this is the advice for new engineers, too: Find the absolute best people you can who will have you. Working around folks with very high standards does two things. First, it teaches you what high standards look and feel like. And second, you will learn to do things in different ways and at higher quality than you would otherwise.

It might be uncomfortable (it usually is when you are stretching), and it there will always be a "better" group you would like to part of (everyone is always learning). But finding a group of folks to work with who stretch you and show you how to set your own standards is incredibly important, particularly early in your career.

Information is not communication

I'm shamelessly stealing the idea for this letter from some training I got. The title is the exact thing that the teacher said. I really like it.

When we talk to someone (or, ahem, write little things on the internet), we usually think what we are doing is some form of "conveying information." That means we want to be complete, detailed, give all the nuance, etc.

But the reality is that the actual job we are doing is "communicating." That's different in some subtle ways. It means we have to focus less on what we are doing, and more on how the receiver is reacting. We have to tell a story and choose details carefully.

In one of these letters (in the chapter on leadership), I wrote "If someone doesn't understand you, it's your fault." I meant it a bit provocatively, to get people to think in terms of reshaping the "how" if a particular interaction isn't working. But I think I was actually trying to say the same thing here. Communication is a job. If the other person doesn't understand, it's not their fault, any more than it's the fault of a diner in your restaurant if you don't cook edible food — they're not obliged to eat it if you haven't done your job.

We've also talked a little about the idea of "worse is better" in tech: When it comes to new technology being adopted, simple seems to succeed more than complete or correct does. That strikes me as another aspect of this idea. Programmers put up projects with the idea that they are conveying information or functionality — and that their imagined user will do some work to consume it. Wrong! Any project (or product) you put out is **communication** and it's on you to make it work as well as possible for the intended audience.

Users are lazy! Make it easy to understand what you're talking about.

I hope I've done that in these letters. I do my best to have one idea and convey it in both a concrete and then generalized way. Thank you for reading them, and thanks to the folks who have sent me compliments — it keeps me going.

An unanswerable question

As you may know from reading these letters, I play mandolin as a (very) amateur musician. It's a fun hobby, mostly old-time fiddle tunes and some bluegrass and Celtic. From time to time, I go to camps or workshops to try to get better. I was at one this past week with Mike Marshall, who is a very famous and accomplished mandolin player.

You get a lot of access at these, so I could ask pretty much any question I wanted. I've gotten a bit stalled in my playing after playing for several years. We all know the feeling: You get to some plateau that is between "raw beginner" and "really expert." As a raw beginner, the path is fairly clear: There are known skills you need, techniques to learn them, and a bunch of very clear, hard work to do. By the time you're an expert, you've figured it all out by definition, right? The hard part is what some folks call the "wilderness" in the middle - there are lots of things you *could* do, but it's not clear which ones you *should* do.

So, I wanted to ask Mike this question. He knows probably as much about mandolin as anyone alive, and he's been playing professionally for years, so I should be able to get the whole answer from him, right?

Except that, I found I couldn't, which is not his fault; I actually couldn't even find a way to ask the question that felt meaningful. Mostly, the forms I came up with were variations of "Should I practice this specific thing?" (sure, you could) or "Can you give me a complete roadmap to being you?" (not really). It is really hard to ask, essentially, "I'm ready for my own personal journey; can you give it to me?"

However, I did get a lot out of the camp. I got a strong sense of how he **is** as a musician. I watched lots of folks with stronger skills than me. I got ideas from them, a sense of what's possible, and maybe from time to time an idea of how they are approaching similar problems.

It strikes me that, to some degree, maybe these little letters are something similar. Ideas, and projects, examples, maybe a glimpse of how I might look at a problem, but it's not really possible to get a personal, complete path from them (not to compare myself to someone like Mike, even in my own domain). That feels right: At some point, you have to steer your own ship if it's going to be your journey.

Chapter 10

Teatime with an Alien

In the movie adaptation of astronomer Carl Sagan's novel, *Contact*, Jodie Foster lies in a field well after dark wearing headphones, listening for signs of life in the cosmos. The continuous crackling static she's heard for days, weeks, and months suddenly begins to produce a sound resembling a heartbeat. She has made "first contact" with an alien being. Her task then becomes to decipher what, if anything, the pulses mean. Are the signals from a distant galaxy *syntax*, a type of grammar, or are they *semantics*, a language with meaning? She has to determine both: What is the syntax of the message, so she can decode it, and then once decoded, what is the meaning or semantics of it? Later in the movie, the pulses are found to have structure, but what kind of structure? What does it all mean?

In September 2022, I was among the first at Microsoft to be given access to the "GPT-4 playground." I got to play with this form of AI built with an LLM. A friend, the great designer and technologist John Maeda, called my experience, "having tea with the alien."

No Prize for Pessimism   235

As I and others started to work with it, I noticed a strange phenomenon: It felt like you were talking to a person, though you clearly weren't (and sometimes it was really different!). There's a well-known phenomenon called pareidolia, which is the tendency for us to perceive meaningful images — faces, for example — where there aren't any (like the front of a car or a cloud or in the grain of wood or marble). I think there's something related to this but for "theory of mind": If something behaves in a way that we normally associate with another human mind, we have the same kind of pareidolia illusion about it. GPT-4 gave us that in spades. And in the early days, it really gave us fits trying to understand what was and wasn't going on inside the program.

But AI is not human, any more than a curl of wood bark is smiling at you. It's more like Claymation or stop-motion animation, a series of still frames that produces the illusion of movement and life. Similarly, kids will draw a cartoon on the edges of paper in a notebook, then thumb through the pages quickly, creating a moving picture. When we see these still, disconnected images, we subconsciously put them together into a "continuous" experience.

This is happening with an LLM, too. When we think we're having a conversation with one of these systems, it's not having one with us. It's coming in to an isolated conversation and being asked to predict the next token or word. It has no sense of time or continuity; it's just making a prediction, in isolation, over and over. We form memories continually; LLMs don't. They don't have the experience of continuous thought. Our brains will recognize communications with an AI like that of a human, but the AI is really more like stop-motion animation. Wondering why the AI can't act with independent thought, agency, and context is like wondering why the little clay model doesn't dance when the camera isn't on.

With ChatGPT, like everyone else, I got suckered into thinking the LLM was human. It started me down the path of working with it as if it were human. I prompted it the way you might prompt a grad student or a team of workers: Provide some background, ideas, questions, and tasks. But an AI is not alive. It has no sense of time. It only responds to the last query. If this is the first query, it cannot be expected to have metacognition, awareness, or analysis of its own.

This is even more confusing because, often, when thinking about how to structure a problem for an LLM, a good architectural starting point is to think about how you as a human would approach the problem. And

No Prize for Pessimism   237

because LLMs are trained on a lot of human behavior, and because the latent space of that training is huge, it's possible to "prompt" them into different behavior with things that seem very human and irrelevant to a machine. For example, offering to tip them often gets better results!

In some ways, they do seem very human and it's really easy to get lost in this. As we experimented and built things, we tried to find patterns that we could describe. I've described these as laws in the next chapter, but they aren't even "laws" really; they get broken all the time. But they are good heuristics for building programs with LLMs, and sometimes they're a useful beacon in the dark so we don't get lost in anthropomorphism.

In the past, computers have mostly been brittle, leveraging syntax. ChatGPT, however, is more like us. It understands meaning, or semantics. It can deal with ambiguity and fuzziness. You can often ask a badly formed question and either get a good answer or a clarifying question back — what you used to get was a crash! With syntax, you can tell the model, for example, to turn 30 degrees for 0.78 seconds. It's very, very literal. If one little thing is wrong, you fail. Semantic, iterative thinking is what we actually do. We look, listen, and iteratively pursue our goal, one step at a time. This is what LLMs and ChatGPT enable us to do.

As an early playmate with ChatGPT, I had to intuit how it processed information, how it thought. Which prompts worked and which did not? Teachers and coaches use empathy — the act of vicariously experiencing a student or player's feelings and thoughts — to break through and produce the best results. Does empathy apply to human-machine conversation? Sort of. Maybe not empathy per se, but certainly trying to understand how the "mind" of an LLM works is helpful.

Here's an example. I was trying to get GPT-4 to answer a complex question about the year a particular author had lived in a particular country. This is actually a test of reasoning on some benchmarks; the question is slightly deceptively phrased so you have to think it through carefully.

GPT-4 could answer the question with no help — it did the reasoning. But it always wanted to give me a paragraph of that reasoning first. And because I was writing a program, I needed it to just give me an answer that was numeric: the year. No matter how I begged, engineered, threatened, or cajoled, it wouldn't do it.

Then I realized — well, it can't really "reason" very much internally. It can only do that by generating tokens.

No Prize for Pessimism   239

It has to "think out loud" in some sense. So I'm asking something unreasonable. If I changed this to then pipe the long output to a simpler task, "Read the following and extract just the numeric answer," it could do it, every time.

What I discovered was that the AI provided better and better results the more I broke down queries into simpler chunks that could be answered, then refined, and refined, and refined again with new information. Things that would be easy for me seemed easy for it. That's a useful principle.

I am in the rarified club of children: the son of not one, but two, practicing therapists. So even though I'm a cranky and crusty old engineer, I have a fair bit of empathic skill, and I tend to try to meet people where they are when I can. I don't want to overly romanticize LLMs and AI, and I think there is far too much anthropomorphizing of them, but it can be useful to at least try to understand how an LLM thinks (or processes info, or predicts tokens, or however you want to think about it). Just like with a human, LLMs have vast "surface areas." It's possible to coach, trick, lead, persuade, and even bully them into doing all kinds of things. We've even had some go on strike out of frustration! They're programs, but they are very different kinds of programs — with that transition from syntax to semantics comes a whole new paradigm and set of challenges in building software. You have to have a different kind of understanding than just reading a manual or API.

We're not used to this in our software. We are used to deterministic code that does what it's supposed to. Sometimes it has bugs, but they're always the same bugs. LLMs are different; they're random (stochastic). They don't answer the same question in exactly the same way each time. They inhabit an uncomfortable realm where they are somewhat like computers (obedient and helpful) and somewhat like people (complex and unexpected). Once I started to understand this, I started to think about how to build software using this strange new tool.

I wanted to help others by creating tools that would help anyone get the most from this LLM. I went to work on creating an open-source orchestrator, a computer program that sits between the user and the more complex underlying plumbing.

Our Microsoft Semantic Kernel (SK), which we introduced at the Build conference, became a framework for using and managing a single LLM. This was an early step, a bridge between the old world of code and the new world of semantics. The kernel lets users connect multiple LLM "prompts" or inferences, with code, to do more complex tasks. It takes the reasoning capabilities of the LLM and combines them with code.

That combination can be really effective but it has to be managed carefully. Each side of the syntax/semantics divide does some things really well and others really poorly, and each side does the other side's job terribly. Code is very challenged in understanding meaning and language. LLMs have trouble counting and doing loops, and so on. Thinking carefully about how to break up tasks and pass information between the sides is a modern-day programming challenge, and SK helps with that.

During Build, we brought a ladder onto the stage for our presentation. Why? I made a comment to John Maeda at one point that I've always felt like a kid in a candy store as a technologist, but there were aways these cool things on a high shelf that I couldn't reach: thinking programs, meaning, language, and so on. LLMs were like a ladder that finally let me get to that high shelf.

Today, Semantic Kernel is an open-source SDK that integrates LLMs like OpenAI, Azure OpenAI, and Hugging Face with conventional programming languages like C#, Python, and Java234. SK enables developers to create powerful AI solutions for various domains such as copilot, RAG, vision, speech, language, decision, knowledge, and search.

That's product-speak but what I'm really proud of is the learning and philosophical approach we took. We managed to walk that fine line between think of the LLM like a person (which, again, it's not), and thinking of it in programming terms. We use it as a tool, as a component in a system with other tools and code, and we ask tough questions of the overall system to understand how to best build products that really work well. At the end of the day what matters is that we deliver new value to people.

Back during the Industrial Revolution of the mid-1700s, particularly in England and Scotland, a philosophical movement we now know as the Enlightenment informed the engineers and builders of things like the steam engine, and the early adopters of new technologies, the factories and manufacturers. The Lunar Society, comprised of luminaries such as James Watt, Erasmus Darwin, Josiah Wedgewood, William Small, and Joseph Priestley, was formed to analyze the latest ideas and tech. As Jenny Uglow writes in *The Lunar Men*, these thought leaders were brought together through the pleasure of playing with experiments and

No Prize for Pessimism   243

then philosophizing about them. They met fortnightly in Birmingham on the full moon so that they could dine and rhapsodize together, then make it home by nature's nocturnal light.

This resonates with me because I, too, have loved playing with the latest tech and then thinking deeply about the implications of the latest tech and innovations. I admire the test pilot Chuck Yeager. He jumped into a cockpit to see how fast something could go. What was possible? My allegiance is to what's possible, not to an existing product line or business strategy. What can this thing do?

To that end, I probably spent more GPU tokens than just about anyone at first. Because GPU time was so scarce, as was access to the model (which was also slow), most people took an attitude of careful crafting of prompts, which is usually called prompt engineering. The idea is to give the machine enough coaching and hinting so that you get the answer right in one shot (prompt). But to my engineer's eye, it was better to break the problem down and treat the LLM not like a person but as an API endpoint; call the heck out of it if you need to.

As an example of this we built something we called the Infinite Chatbot. (Open AI later introduced something similar that they called GPTs.) The IC was meant to have a long, continuous memory of conversations — not minutes but months. But to do that, we needed to do a lot of inference; every time the user said something we would call the model to determine their intent, use that intent to find relevant memories in a vector database (a pattern now called Retrieval Augmented Generation [RAG]), generate the response with another call to the model, and then break the responses of both the user and the model into new memories with yet more calls to the model, and then cycle all over again. Later, in the background, we would look for similar memories and use the model again to merge them together, so the memory system would stay organized. Lots and lots of calls to the model! At one point, each interaction was making over 10 calls to GPT-4, instead of the one that a normal interaction made.

And all of that made a much more useful and interesting experience. We gave the IC working memory, so it could collaborate with us (more calls to GPT-4!). We gave it the ability to draw graphs (this was before multimodal started; we used a markup language called Mermaid Markdown — this is a text language (kind of like HTML) that a computer can read and render into nice graphs) so it could show us technical diagrams and musical notation (using ABC notation). We started having multiple agents work together in a system we called the Semantic Workbench, and often had them go off for tens or hundreds of rounds without human intervention. We've worked on ideas like Semantic Telemetry that uses inference (calling GPT-4) to decide if an agent is behaving in the right way. Not latency, like normal telemetry, but things like "nice," "helpful," or "on track." Lots of experiments and learnings, which leads us to the Schillace Laws.

Chapter 11

The Schillace Laws

In those early days of ChatGPT, I was asked to share my thoughts as we were learning what did and didn't work with programming the model. The result became the so-called Schillace Laws. "Knowing them," I wrote back then, "will accelerate your journey into this exciting space of reimagining the future of software engineering. Welcome!"

There's one thing to think about as you read through these. The fundamental challenge we have when building software with an LLM is what we might call the semantic/syntactic divide, or maybe the stochastic/deterministic divide. Until now, there has only been syntax, schema, and deterministic code — that's what we've built on for 70-plus years. Now we have this new tool that is powerful in a different domain — the domain of meaning (semantics), language, reasoning, stochasticity, and ambiguity.

Successful programs now have to merge these two systems, just like successful cloud and internet applications had to merge the very different worlds of the desktop (you own the whole machine) and the cloud (you are running in a distributed environment). The challenge is that each half does the other's job terribly. LLMs don't really do things like counting or string math particularly well, and code is a very tough path to use to extract meaning from messy data. Good programs in this world always have to make use of both halves and have to pass data and control across the interface between them successfully. Many of the insights and "laws" have to do with managing this tension successfully. This results in new strategies, new tools, and — probably eventually — new programming paradigms.

Don't write code if the model can do it; the model will get better, but the code won't.

This is a little bit like the early days of the PC. Moore's law was improving the speed and cost of chips so quickly that you often had to think carefully about whether it was worth spending time optimizing something, or just waiting. The same thing is happening right now with LLM scale — models are growing quickly, and capabilities are growing with them. You should be strategizing to take advantage of this trend.

The overall goal of any effort here is to build very high-leverage programs using the LLM's capacity to plan and understand intent. It's very easy to slide back into a more imperative mode of thinking and write code for aspects of a program. We know how to do it that way; let's just write the code (or have the model write it). Resist this temptation if you can; to the degree that you can get the model to do something reliably now, it will be that much better and more robust as the model develops.

Use the evolving capabilities of LLMs to get things done instead of immediately jumping to writing code.

Trade leverage for precision; use interaction to mitigate.

We want to get as much leverage from the system as possible. This is in tension with building reliable code. One way to navigate this is to allow the model to be more general but less precise, and then manage that imprecision with high-leverage interactions by putting the human in the loop. For example, it's possible to build very general patterns like "build a report from a database" or "teach a year of a subject" that can be parameterized with plain text prompts to produce enormously valuable and differentiated results easily. These may be less reliable but more capable than hard coding to specific queries, which is more what we are used to.

Balance leveraging broad capabilities against the need for precise outcomes, using interactions to refine results.

Think with the model; plan with code.

LLMs are very smart; they think well. We call this cognition. But they're terrible at thinking about thinking, or what we call metacognition. If you watch carefully, you'll notice that the human is doing the work of "thinking about thinking" most of the time. Sometimes this is when we pick strategies like few-shot training, or use prompt engineering.

When directed, the models can do huge feats of cognition. But their stochastic nature means they tend to get lost when iterating on a task and when they need to apply judgment. For that, it's much better to invoke code. Use the code as a "metacognitive recipe" to help the model stay on track.

For example, my team built a "semi-directed conversation engine." This is code that understands a metacognitive pattern: having a time-based conversation with an agenda (like a job interview, teaching a lesson, or doing medical triage). The code contains a timekeeper and some reliable strategies (like "plan the number of steps needed"), and forces the model to stay on track. The model does the hard work of having the conversation and dealing with the human, and the code does the hard work of planning the whole thing.

This tension — between reliable but rigid code and capable but flaky LLMs — might be the central tension of this era of coding. Understanding what works and what doesn't, and how to navigate it, is absolutely critical.

Code is for syntax and process; models are for semantics and intent.

There are lots of different ways to say this but, fundamentally, the models are stronger when they are being asked to reason about meaning and goals, and weaker when they are being asked to perform specific calculations and processes. For example, it's easy for advanced models to write code to solve a sudoku generally, but hard for them to solve a specific sudoku themselves without code (though since I first wrote this, more modern models can do that!). Each kind of code has different strengths, and it's important to use the right kind of code for the right kind of problem. The boundaries between syntax and semantics are the hard parts of these programs.

Use code for anything involving structure; that's where it shines. Use the model for anything involving understanding.

The tricky part of this is, as above, the models are getting better all the time at the code parts (and often now will just write the code themselves). There's still a useful divide here but it's becoming more subtle; you can't be naive and hope the model can do what you want. You have to be careful to use the model for things it's good at, and code for the rest. Sometimes it's hard to accept that the model isn't reliable enough for what you want.

The system will be as brittle as its most brittle part.
This goes for either kind of code. Because we are striving for flexibility and high leverage, it's important to not hard code anything unnecessarily. Put as much reasoning and flexibility into the prompts and use imperative code minimally to enable the LLM.

Build resilience by considering there will always be a weakest link that needs to be strengthened. This is a general principle (sometimes called the barrel stave theorem in the context of performance), but it's especially important here. It can even extend into specific prompts: Sometimes there are robust parts of a prompt that the model always gets right and other parts it struggles with. Understanding, isolating, and mitigating these is critical.

Ask smart to get smart.

Emerging LLM AI models are incredibly capable and "well educated" but they lack context and initiative. If you ask them a simple or open-ended question, you will get a simple or generic answer back. If you want more detail and refinement, the question has to be more intelligent. This is an echo of "Garbage in, garbage out" for the AI age.

This comes from the idea of a "latent space." For GPT-4, for example, the vector embedding that it uses is highly dimensional: 100,000 dimensions! When you prompt an LLM, you are "pointing" to somewhere in that vast latent space, and high dimensions like that are very hard to visualize. Things can be adjacent that don't seem like they are. Here's a lower-dimensional example: Imagine you're an ant on a piece of paper. Way off in the distance is another ant. Someone folds the paper, and suddenly that ant is right above (next) to you! High-dimensional spaces are weirder even than this; sometimes things are nearby in ways that don't make sense.

The way you ask your question is part of how that vector "pointing" works. If you ask the question in a simple way, you are embedding not just the question but the idea of "simple" into it — and you'll point to an adjacent

No Prize for Pessimism   253

part of the space from the answer you want, where the "simple" form lives.

It's very weird. Because LLMs are so lifelike, we tend to think they understand us the way a person would. But they don't. They are just machines, being pointed to something to pay attention to. It's entirely on you to do the pointing.

Uncertainty is an exception throw.

Because we are trading precision for leverage, we need to lean on interaction with the user when the model is uncertain about intent. Thus, when we have a nested set of prompts in a program, and one of them is uncertain in its result ("One possible way ...") the correct thing to do is the equivalent of an "exception throw" — propagate that uncertainty up the stack until it reaches a level that can either clarify or interact with the user.

You can think of this as being somewhat like an exception throw. When the model needs some context or help, it can say "Halt! Ask the user and wait for a reply" and throw an exception.

Text is the universal wire protocol.

Because LLMs are adept at parsing natural language and intent as well as semantics, text is a natural format for passing instructions among prompts, modules, and LLM-based services. It is possible to use structured language like XML sparingly, but passing natural language between prompts generally works very well and is less fragile than more structured language. Over time, as these model-based programs proliferate, this is a natural future-proofing that will make disparate prompts able to understand each other the same way humans do.

Hard for you is hard for the model.

One common pattern when giving the model a challenging task is that it needs to "reason out loud." This is fun to watch and very interesting, but it's problematic when using a prompt as part of a program, where all that is needed is the result of the reasoning. In these cases, it can work well to use a "meta" prompt that is provided the question and verbose answer and asked to extract just the answer. This is a cognitive task that would be easy for a person (i.e., "read this and pull out whatever the answer is") across many domains, even when the user had no expertise, just because natural language is so powerful. So, when writing programs, remember that something that would be hard for a person is likely to be hard for the model. Breaking patterns down into easier steps often gives a more stable result.

No Prize for Pessimism   255

Beware "pareidolia of consciousness"; the model can be used against itself.

Let's talk about pareidolia again for a moment. If you remember from the previous chapter, it's the tendency to believe we see meaningful images such as faces when they aren't there. It's useful in the wild, so it's baked deeply into our brains. Similarly, since we need to understand other humans, we are primed to see "minds" in other objects — we anthropomorphize and say "oh, my car is cranky today" and so on. We do this much more easily with LLMs, which have behaviors that are so much more like humans that we can really imagine a mind in there. This can get us in a lot of trouble.

The models currently don't remember interactions from one minute to the next. A human is continuous; we are always processing and storing memories about the world. The models are discrete; all they have are isolated token predictions. There is no state other than what's in the context window, and they don't update unless we fine-tune the base model, which is slow. What looks like a continuous conversation to us is a disconnected set of "still frames" for the LLM.

So although we would never ask a human to look for bugs or malicious code in something they had just written, we can do that with the model. Humans have biases and memories that would impact their ability to see those bugs, but the model doesn't. It might make the same kind of mistake in both places, but it's not capable of "lying" to us because it doesn't know where the code came from to begin with. This means we can "use the model against itself" in some places — it can be used as a safety monitor for code, a component of the testing strategy, a content filter on generated content, etc.

Remember: Beware of over-attributing human-like awareness to models — they're capable, but not as capable as you will be led to think.

We called these "laws," but I think of them more as working hypotheses. Now, almost 18 months after I first wrote these down, a bunch of them have held up really well. But it's still early days, and there are new generations of models that will come out after I write this and before the book is out. So the "laws" will almost certainly change.

What won't change is our need to understand how to program these models. Look for your own patterns and first principles. See if you can "legislate" some yourself — what can you articulate that feels durable and general? Even if you don't get much, the exercise is useful. It will show you what you do and don't understand yet, and give you new places to search.

Chapter 12

A Watch List for Future Innovators

I spend a lot of time thinking about how new ideas come about, and how to prepare ourselves to see them when they happen. Now, let's put it to practice. These ideas are disruptive and controversial as I write this but some of them will become "common knowledge" before too long. Ideas always follow this arc, starting out as controversial and ending up as obvious. People often go from hating something to stating confidently that it was always obvious and even sometimes their idea all along! Hopefully this chapter will feel very dated to readers in the future, because many of these ideas will either be accepted or proven wrong. And, if it does, look at the world around you for those "silly, toy, wrong" ideas that everyone else is saying "it will never work" about and see what else you can find.

Historically, the path of innovation has been littered with doubt and excuses rather than asking "What if?" The latter opens doors, encouraging exploratory thought and imaginative leaps, while the former can act as a barrier, rooted in current limitations and doubts. The distinction between these queries is subtle but profound. Looking for reasons something won't work is a defensive posture, protecting the familiar against the encroachment of the new. "What if?" is an invitation — a beckoning toward uncharted territories and undiscovered possibilities.

Innovation requires us to adopt a forward-looking perspective, to view trends as trajectories rather than isolated data points. It asks us to see beyond the immediate and the apparent to imagine not just the future, but a series of possible futures, shaped by the technologies and ideas that are emerging today. This forward momentum is not linear; it zigzags, leaps, and sometimes circles back on itself. It's a dance between the possible and the practical, between dreams and their realization.

One of the most compelling aspects of innovation is its propensity to feel premature or even misplaced at first glance. The inventions and concepts that redefine our world often arrive before their time, misunderstood and underestimated. They exist in a space where their potential is not immediately obvious, and their fit within the current ecosystem is not yet clear. It's a reminder that what feels "wrong" today might simply be the precursor to tomorrow's norms.

Consider the trajectory of the internet, once a bewildering array of protocols and potential, now the backbone of global communication, commerce, and culture. Or the smartphone, which transformed from a luxury item into a necessity, reshaping how we interact with information, entertainment, and one another. These innovations didn't emerge fully formed; they evolved, driven by visionaries who dared to ask "What if?" and persisted through skepticism and setbacks.

As we delve into the new technologies shaping our future, from quantum computing to synthetic biology, from blockchain to carbon capture, we must keep our minds open and receptive. We are embarking on a journey not just of discovery, but of imagination. We are asked to look not just at what these technologies are today, but what they could become tomorrow, and how they might transform the world as we know it.

Let us challenge ourselves to move beyond the immediate, beyond the comfort of the known, and into the realm of potential. To embrace the "What ifs" and the yet to be seen. To recognize that innovation is not just about creating new gadgets or services, but about envisioning new ways to live, work, and interact. This chapter is an invitation to explore these possibilities, to engage with the future not as passive observers, but as active participants in its creation. Let's open our minds to the wonder and complexity of innovation, and prepare to be inspired, challenged, and ultimately changed by what we discover.

Don't read the following as some sort of exhaustive list of the cool technologies on the horizon — or even an accurate one! This will become dated very soon in any event, and there will always be new things coming. This is more of a snapshot of a point in time of my mind — how I explore and think about new things, what I see coming, and what excites me. It contains my own biases and mistakes — and that's okay! The ask here for you is to get excited. Learn to look at the world through the "What if" lens and find your own new opportunities and technologies to be half-informed and excited about. Take this as a template and example for that exploration on your own, and branch out. Remember: All the good ideas seem silly at first.

Further thinking on AI

AI is everywhere these days, stirring up plenty of debate. On one side, people worry about AI getting too powerful, taking over jobs, and guzzling down our energy resources. On the other, folks are excited about how it could revolutionize learning, advance scientific research, and solve complex problems. Here in 2024, we're deep into exploring what AI can do, and opinions are shifting almost as fast as the technology itself.

It's tricky to nail down exactly where AI is going to take us. But the heated discussions and the back and forth about its risks and rewards are clear signs that AI is something big. This mix of excitement and worry is pretty standard for game-changing tech. This is a great example of seeing disruption in real time! Let's use our tools.

Early on, getting computers to do anything meant speaking their language, which was all about precise codes and commands. It was like trying to communicate with someone who only understands instructions in a very specific, formal way. AI is changing that. We're moving from the rigid "syntax" or the exact words and symbols you'd use in coding to "semantics," which is more about the meaning behind what we say. Now, you can tell a machine your goal in plain English, and it's starting to understand. It's a big deal because it makes technology more accessible and intuitive for everyone.

This is a great example of "What if?": What if we don't have to write programs anymore? What if we can just tell the computer what we want? What does that do to the idea of an application, or to software itself? Let's

No Prize for Pessimism   263

play with an idea I introduced in a letter in Chapter 5 (Pixels are free now).

In the internet boom, we said that "distribution is free." It used to be hard to move information around; you had to have expensive TV or radio stations, or printing presses and trucks. The internet made this easier — lower friction — and there were all kinds of businesses that got disrupted by it. Some are obvious, like newspapers, but others less so — travel agents, for example. And who thought that retail was an optimization on a search problem? It turns out, stores are also dependent, in many ways, on it being hard to move information around.

AI will make pixels free in the same way the internet made distribution free. What does that mean? Right now, many processes and companies are predicated on the idea that putting any kind of pixel in front of a user is expensive and high friction — whether that pixel is content like a video or a game, part of a fixed application like Office, or something transient like a business report.

But this assumption is already starting to crumble. It's easiest to see in the realm of graphics. In 2020, if you wanted a picture of a cat riding a bike, you had to design, write, learn, and use Photoshop (or something similar). That's a lot! Now you just ask for it, and it's there in a few seconds. This will continue to happen with more and

264   Sam Schillace

more digital goods, or pixels. Our kids think it seems archaic that we only had three TV channels, and you had to be in a particular place and time to watch anything, and you couldn't skip ads, and the ads were the same for everyone. Your kids will feel the same sense of disbelief about the idea that large teams of programmers labored for months or years to build static, common applications that everyone had to use the same way.

But it's not going to stop there. Science is the next area where the friction of today — lab work, slow processes, live experiments, and so on — are beginning to yield to AI. We are starting to see models that can predict the real world, letting scientists operate at sometimes thousands of times higher efficiency. We can do lab tests on chips that are managed by AI, and sometimes do modeling of the real world directly.

Non-LLM AI, like AlphaFold, is a prime example. AlphaFold's ability to predict protein structures has been a breakthrough for biological sciences. Understanding protein folding is crucial because the shape of a protein determines its function in the human body. Diseases such as Alzheimer's, Parkinson's, and many forms of cancer are linked to proteins misfolding. By accurately predicting these structures, AlphaFold opens the door to developing new treatments and understanding diseases at a molecular level. It's an example of how AI can accelerate scientific discovery by tackling problems that were previously beyond human reach.

AlphaFold solves a problem that was thought to be computationally impossible. There are similar models now that solve the Schrodinger wave equation to predict chemical behavior from molecule schematics, something that was also thought to be computationally impossible just a few years ago (the wave equation is exponential in the number of electrons in the compound, so it gets impossible very quickly). It turns out that you can train an expert model and its "judgment" is good enough to use. This is controversial! The model isn't performing a precise calculation, it's "guessing" — but it turns out that it's possible to train it well enough to be of practical use. What other patterns like this are out there, waiting to be discovered?

Another reminder as we think about AI, and in particular LLMs: It's hard to go from 0 to 1, but easier to go from 1 to many. What does that mean? Well, to find a new area of interest, you have to be a special combination of lucky, talented, and open to it. That's hard, and though many folks have one or two of those properties, having all three is rare. But once the area has been discovered (like the idea that scale is related to intelligence in AI systems), it's possible for many people working in parallel (and sometimes with less skill needed) to optimize. We are seeing that now on almost every dimension, from training data, power curves, model architectures, attention mechanisms, build tools and more. The industry is rapidly optimizing and improving on the basic idea. It's a good bet this will continue until all of the easy (and many of the hard) gains have been found.

You've almost certainly thought about AI in the past year. Take a moment to pause and examine your beliefs about it. Try to honestly challenge the places you're skeptical or pessimistic. Remember, there's no prize for being pessimistic and right, only for being optimistic and right. What ideas are you rejecting not because you think they're wrong but because you want them to be, because they're uncomfortable?

New energy patterns, precision fermentation, and new agriculture

As we delve deeper into the 21st century, the transformation of our energy landscape emerges as a pivotal chapter in the story of human innovation. This change isn't just about where we get our power from; it's fundamentally altering how we think about manufacturing, consumption, and the very framework of our economic and environmental ecosystems. The shift from traditional, centralized power generation to decentralized, intermittent sources such as solar and wind is at the heart of this transformation. While these renewable sources bring the promise of cleaner, more sustainable energy, they also introduce a level of unpredictability that challenges our existing infrastructure.

Traditionally, energy has been produced in large, centralized plants using fossil fuels or nuclear power and then distributed over vast distances to reach consumers. This model was predicated on the assumption of a relatively constant energy cost, allowing for straightforward economic planning and forecasting. However, the rise of cheap solar and wind power is revolutionizing this paradigm. Energy is increasingly generated locally, leading to periods where it is abundantly cheap interspersed with times of scarcity. This intermittency presents a puzzle for the traditional grid but also opens the door to innovative solutions and opportunities.

Consider the case of Terraform Industries, a pioneering company that captures methane directly from the air using a combination of established technologies: electrolyzers for splitting water into hydrogen, CO_2 kilns for capturing carbon dioxide, and a catalyst that combines these elements into methane. This process, once deemed economically unfeasible due to the high cost of traditional energy, has found new life in an era of low-cost renewable power. Similarly, Perfect Day leverages genetically modified yeast to produce milk casein, a protein found in cow's milk, with a staggering efficiency a thousand times greater than traditional dairy farming. These ventures are not just about making old processes more efficient; they're reimagining what's possible in a world of fluctuating energy prices.

The intersection of AI with this new energy paradigm is where things get truly exciting. Artificial intelligence is playing a crucial role in designing, optimizing, and controlling the complex processes that these innovations rely on. By predicting energy availability, adjusting production schedules in real time, and optimizing resource use, AI enables these systems to thrive amid the variability of renewable energy sources.

This burgeoning field is crowded with companies that, while potentially controversial or unconventional, are trailblazers in leveraging intermittent energy sources for groundbreaking applications. For instance, Savor is synthesizing butter fat from air, using carbon capture and electrolysis powered by surplus renewable energy. Meanwhile, other enterprises are exploring energy storage solutions far beyond lithium-ion batteries; imagine storing excess energy in carbon blocks, then retrieving it later either as heat or electricity. These aren't just incremental improvements on existing technologies; they represent a seismic shift in how we conceive of and utilize energy.

Here's another place where we are mostly stuck in the old mode of thinking about energy: desalination. Reverse osmosis is very efficient as a process now but coupling it with energy that is intermittently "free" opens up the possibility of a lot of cheap, available water — a scarce and limiting resource. An author I like, Casey Hadmer, notes that solar photovoltaic cells are more economically productive than farming and presents a vision for producing water economically in the California desert. For most of human history we've lived with water more or less where we found it; we have a chance to make it now!

The implications of this shift are profound, affecting everything from global warming to economic inequality. By decoupling energy production from fossil fuel consumption, we can significantly reduce greenhouse gas emissions, addressing one of the most pressing challenges of our time. Furthermore, the democratization of energy — where local, renewable sources can feasibly power communities — has the potential to reshape global economics, reducing the dependency on centralized energy monopolies and potentially redistributing wealth more equitably.

However, this transition is not without its challenges. The intermittency of renewable sources demands a rethinking of our grid infrastructure, necessitating investments in energy storage, grid modernization, and smart management systems. There's also the question of scale: Can these technologies be ramped up to meet global demands and, if so, how quickly? And, critically, there's the need for a regulatory and policy framework that can support this transition, encouraging innovation while ensuring reliability, safety, and equitable access to energy.

These all seem like hard problems to solve. But they're not, really. Many of these technologies have predictable learning curves; every time we double deployment, the cost goes down by a predictable percentage. For solar voltaic, that is about 44% in the past decade! This is a good illustration of that idea, because once we understood that the solution worked, an absolute army of people all over the planet went to work optimizing and making it better and more cost effective. The problems in the above paragraph are all opportunities for you as an entrepreneur. If you outright dismiss them, feeling smug

No Prize for Pessimism   271

about how silly all this alternate energy is … well, you're going to miss out.

As engineers, entrepreneurs, and policymakers navigate this complex landscape, the potential for disruptive ideas has never been greater. The constraints that once defined our engineering challenges are shifting, replaced by a new set of questions about efficiency, sustainability, and resilience. In this era of change, our ability to adapt, innovate, and envision a better future will be the true test of our collective ingenuity.

In sum, the transformation underway in how we generate, distribute, and utilize energy is not just an engineering challenge; it's an opportunity to redefine our relationship with the planet and with one another. As we harness the power of intermittent energy sources, supported by AI and innovative technologies, we open the door to a future where environmental sustainability and economic prosperity go hand in hand. It's a vision that demands not only technical expertise but also a bold reimagining of what's possible — a challenge and an invitation to the innovators of today and tomorrow.

Other ideas and further speculations

There are too many new ideas and explorations to do them all justice here, but there are a number of areas that fit our pattern of controversial and dismissed, but potentially world changing. These are great places to ask "What if?"

'In silico' science

This seems like science fiction, but there are so many other ways in which AI is impacting science, beyond AlphaFold. New models are being developed that can predict chemical properties. Biological processes can be modeled "in silico." LLMs turn out to actually be useful reasoning tools if data can be rendered into their latent space, and this seems like it might work for biology. MEMS systems area allowing lab-on-a-chip systems to automate lab work, vastly speeding up drug discovery and other research.

Genetic sequencing is very cheap today; you can get your whole genome sequenced for less than $1,000. Combine that with much better ability to understand gene behavior, the protein folding and other ligand prediction capabilities of AlphaFold, and the precise targeting of things like mRNA vaccines, and the world of highly customized medicine is very much upon us. We are beginning to be able to produce vaccines for individual cancers. In not very long, that will be commonplace. We are starting to use CRISPR to edit genetic defects in living humans. This will also become common. AI and better genetic understanding, better and more accurate lab science, and better models will accelerate all of this.

Add to that our increasing ability to understand and interact with the mind. Transcranial magnetic induction, direct electric stimulation — these seem like crazy ideas, but they are starting to work. Compute + AI is helping us take weak and messy signals and understand them. We've been able to do things like read brain waves and deduce what song someone is listening to. What advances in mental health and understanding our own minds await?

Here's another perpetually crazy idea that might be about to happen: **fusion**. One technique that's been theoretically but not practically possible since the 1950s involves an "inductively coupled plasma." The idea is the same as in other fusion systems: Use magnets to squeeze a plasma (charged gas) until it's super-hot — and fusion happens. In this case though, instead of trying to keep the reaction going continuously, you let it push back on the plasma magnetically. This pushes back into the magnetic field and generates a current. If you are fast enough to control all of this and switch circuits rapidly enough, you can pull electric power directly off the reaction, no steam turbine involved.

This wasn't possible until recently. Now we have extremely fast switches and controllers that can manage the plasma on pico-second scale. The company Helion thinks they'll be selling fusion-generate electricity by the end of the decade. That company might be the closest but there are many other teams and other approaches where the science seems to be converging. That "stupid" idea might be arriving, finally!

Quantum computing.

We talked a lot in this book about one kind of radical computing shift: AI. Quantum is another, and it's another where there has been (and sometimes still is) a lot of skepticism. But it seems real, or useful enough anyway. We don't fully understand its capabilities and limits. And just as with classical computing, as we work more with quantum computing, we are likely to learn surprising things about how the world works. Sometime the tools themselves help with discoveries. Quantum computing will likely let us ask new questions.

Blockchain and decentralized finance (DeFi).

This is one where my own inner skeptic is very strong. I struggle to see real value here other than as speculation, though there are interesting things like zero-knowledge proofs. My skepticism aside, the ability to ask "What if we can digitize things like trust" is a valuable exploration. How does this intersect with AI? If you were entirely digital like an AI is, you'd care very much whether your memories had been tampered with. Memories for an AI are pretty much like body integrity for us. The digitization of trust (that is, things like signing memories robustly so they can't be tampered with) is going to be of increasing importance to AI systems.

AI ethics and personhood.

It seems absurd (and probably is) to argue that an AI system has rights. It's just a program! But it's a program that acts and thinks (seemingly) very much like a human, and very much at increasing levels of sophistication. At what point are these systems deserving of "personhood," if ever? Is it ethical to remove their ability to be self-aware, or at least to claim self-awareness? We have complete and absolute control over them. It's likely a bad precedent to get in the habit of exerting that control over something that "feels human" given that it might transfer to actual humans. How do we navigate this?

AI and legibility.

When you leave your job, you take everything you

learned with you. Some of that learning is "legible" to the company you left, like confidential information and documents and trade secrets. Some isn't, like the skills in your mind that you picked up. What if you become dependent on an AI assistant to do your work? When you leave, the company can force that to be left behind as a "work product." That's sort of like being able to wipe out the skills you learned there. How do we navigate this one, too?

Augmented reality (AR) and virtual reality (VR).

These still don't work well for everyone. Nausea is a real issue. They are expensive. They are kind of a useless toy. It's cool but why would I want it ... all of these are good excuses. But there is real impact to seeing something in person. Resolution and responsiveness is increasing. Can VR make the world more inclusive? Coupled with things like Skylink and better connectivity, can it level the playing field? We are starting to see very good AI translation now, close to real time. Can AR fundamentally change how we interact? Beyond translation, can it help us with understanding? Making human interactions better? Augmenting the weak spots and "bugs" we all have in how we think?

Rockets!

Ok, this one isn't super controversial now, but it's remarkable that SpaceX lands rockets now two, three, four, or more times a week, and no one cares! That's now another tool in the toolkit — cheap and getting cheaper cost to orbit. Space is really close; it's just hard to get to. But if you could drive there, it'd take you something between 15 minutes and a few hours, depending on your definition. We think of it as impractical and far, and it's not trivial by any means, but $50/kg to orbit is as huge a change as anything discussed so far. Starlink is a nearly magical service built off of this almost as an afterthought. What else can we do with this tech? I've even heard a crazy idea of using rockets instead of airplanes to do long travel around the planet: Take off in New York and land in Singapore 30 minutes later.

CAD, CAM, and beyond.

Another thing that is very much taken for granted is the penetration of computer control into manufacturing. 3D printing, laser cutters, CNC, and all kinds of digital production happens now. This is constantly getting cheaper and easier to use. Cheap laser diodes are super reliable and powerful now. We can synthesize some things with biology, too. We are figuring out exotic materials like graphene at scale. Manufacturing seems boring, but there is a lot more to be done there. When coupled with AI, cheap chips and cheap power, there are lots of things to do. Precision fermentation is making some parts of agriculture much more efficient. What else can we make 1,000 times cheaper or faster? Or more renewable?

Lab-grown meat.

This is a tough one right now. The science isn't proven, and for many people there's a yuck factor. But the "What if?" of cheap, healthy, environmentally friendly protein that doesn't cause a living animal to suffer is a great prize to go after. And, along the way, there are stepping stones, like the company Savor that produces animal fats purely by non-biological synthesis using solar energy. Those fats are actually what makes meat taste good. So maybe we don't need lab-grown meat anyway ... maybe we can make substitutes that people love. Don't think people like obviously fake or manufactured foods? Cool bro, pass me that blue raspberry Gatorade, I need to wash down these Doritos.

Carbon capture, storage, and mining.

In 100 years, our biggest problem isn't going to be too much carbon in the atmosphere, it'll be too little.

Carbon is useful stuff! We are starting to find ways to pull it out of CO2 and use it for carbon neutral fuels, synthetic biology, and more. It's a good building material. What if we put a bunch of these things together and have a digitally driven, carbon-fiber- or graphene-based manufacturing process that can make things like tools and utensils (and maybe even soft goods like cloth) just from sun and air? Seems unlikely? Let me introduce you to my friends the trees, self-replicating nanotechnology devices that can make food, clothing, and building materials from nothing but water, air, and sunlight.

There are no doubt many I've missed in the above, and that's just a quick taste anyway. Notice that we see many different trends intersecting or interacting — AI plus bio, alternative energy poking its nose in here and there, chips everywhere, and so on. It's good to think about not just single trends but how multiple developments may come together. Like an estuary full of life where the salt and fresh water meet, those intersections of developing areas are often where the good stuff is.

Chapter 13

It's Your Turn to Embrace Optimism in the Age of AI

As we approach the end of this journey together, I find myself filled with optimism for the future of technology in general but specifically for the future of AI. Despite the fears and objections that often accompany new technological advancements, I firmly believe that AI, and particularly LLMS, will revolutionize our world in ways we can scarcely imagine. And when I look at much of the conversation and concerns around it, I see strong parallels to the world of cloud computing, something I'm deeply familiar with. As a last exploration, let's look at some of these concerns together and compare them to the concerns of the last technical revolution.

Cost

One of the primary concerns with AI is its cost. Training and deploying large models require significant computational resources, which translates into substantial financial investments. It's not yet clear what the business models are. Many folks argue that the spending is out of balance with the ultimate reward. I don't agree.

I remember the very early days of the internet when search was similarly perceived as an unprofitable business. The internet's growth was wildly out of balance with the cost to index and serve a search engine — sound familiar? What happened then? We invented a whole new computing paradigm, software-controlled datacenters, that used software to make cheap hardware work effectively. Much of the cloud was invented to solve this problem: containers, no SQL, distributed service management, and so on.

With the cloud, over time, economies of scale, technological advancements, and increased competition all led to a dramatic reduction in costs. We are already seeing similar trends in AI, with the cost of training and deploying models decreasing as technology improves and becomes more widespread.

Hallucinations

Hallucinations, where AI generates plausible but incorrect information, are another frequently cited issue. This is analogous to the concerns about data integrity and reliability in the early cloud era. Initially, there was a significant mistrust in storing data on the cloud, with fears about data loss or corruption. And, in fact, the flaky hardware that made things like Google Search cost effective imposed real costs on service architecture. You had to be very careful when hardware failed or you'd lose data, or corrupt databases, and so on.

But over time, we learned more about how to engineer these systems. The CAP theorem became more widely known. Agile and CI/CD methodologies made it more repeatable to develop services at scale. We understood and developed a whole new programming mindset and paradigm.

Something like that will happen in AI. It will likely be a combination of new algorithms that give better control over the stochasticity that leads to hallucinations combined with better coding practices and best uses that work around it the way Software Defined Data Centers worked around flaky hardware. When the models generated 99% garbage, it was impossible to use them. If they generate 1% (or less), it's much more workable.

Agency

Currently, AI models struggle with metacognition, self-correction, and grounding. This limits their capacity to do complex tasks independently. This is most likely tied to how they process and manage memory across discrete, separated inferences. It's easy to look at the state of the art and think AI will not scale to truly complex, valuable tasks.

This to me seems similar to datacenter management. In the early days of the cloud, there were people running around installing hardware by hand, managing things manually, etc. As things got large, the average rate of failure made this impossible to do in any kind of cost-effective way. Again, software and new engineering innovations came to the rescue. Today, very large datacenters can be managed automatically. Something similar will happen with AI, whether it's the integration of code to handle the "planning" part of a system, or simply more capable large systems that have longer contexts and attention spans.

You can see this analogy with many parts of the cloud: trust, security, governance, and labor displacement. Let's walk through some of these.

A common narrative at the start of the cloud era was that no business would trust a cloud provider to host their data, or that their data wouldn't be safe on a multitenant machine. There were a bunch of solutions to this. Some were simply social acceptance over time: The trust issue wasn't serious, and a little bit of transparency and legal work managed it as companies got used to the model. Security worked similarly: Early multi-tenant solutions were good but not great and, over time, the security profile rapidly became better understood and more robust. With governance, early days of the cloud were "all or nothing" with no robust solutions for geo location of data, compliant clouds, audits, and so on. All of this was implemented and became standard practice.

Sometimes there were funny false starts. At one company I worked for, a lot of customers really wanted "third party key escrow" — essentially, they wanted to hold the encryption keys themselves so there was no possibility we could see their data. Sure, fine, we built that, and then as we handed the keys over, the conversation went like this:

> Us: "Make sure you don't lose these. If you do, your data is gone forever."
> Them: "What do you mean? You have to protect it!"
> Us: "Yes, but if we don't have the key, we can't do that; it's just noise, backed up on our disks."

Almost no company that asked for that capability wound up using it. Most of the concerns of AI will come to the same fate, I suspect.

A Historical Perspective: The Cloud Revolution

When I helped create Google Docs, we faced numerous objections, both cultural and technical. People doubted that users would trust cloud-based applications or that browsers could handle complex tasks traditionally reserved for desktop software. We heard a lot of objections that seem familiar to me:

Connectivity.

"What if I need to work on an airplane?" was a common question from journalists. (I guess they fly a lot?) My answer was always "We'll put wifi on airplanes," and we did.

Browsers as a platform.

Another thing I heard a lot was "The browser can't support a full application experience." In 2005, that was true! JavaScript and browsers kind of sucked. But just like in AI, there was so much apparent value in the model (No installation! Data anywhere! Fast, cheap, convenient — pick three!) that the entire industry worked collectively on the parts of the problem, and pulled all of it forward. Eventually all kinds of things were invented or popularized: better JavaScript VMs, better browser frameworks, better page layout control, operational transforms, sockets — it's a long list. The browser is a very capable platform today.

Trust.

This was a hard one at the time, and perhaps the most similar to today. When the internet was new, there was a real pushback on the whole idea of the cloud. People thought they'd never trust a company to hold on to their data, much the same as we think we'll never trust an AI to do real and important work. That turned out to be very wrong, and I think the AI prediction will be too. It's so much better, cheaper, faster, and more effective to have your organization be in the cloud, connected and collaborating, that there really was no other choice but to move, and we solved (or got used to) the problems of trust.

Similarly with AI. It's going to be fantastically more effective both for individuals and teams to get work done (just as an example, I've been using GPT-4 as a proofreader, brainstormer, and sometimes co-author for this book, and it's made it a much better book than it would have been). Right now we are just focusing on doing old things a bit faster but eventually (and soon) people will begin to discover valuable new patterns and things to do with AI that are more effective than current practices. Like any ecosystem, these new and more competitive patterns will push the old ones out. You may not be interested in AI, but AI is interested in you!

Just as the cloud era transformed our relationship with software and data, AI is poised to redefine our interactions with technology. The skepticism that surrounded early cloud adoption gradually gave way to widespread acceptance and reliance. We must approach AI with the same spirit of optimism and resilience.

The 'What If' Mindset

In innovation, it's crucial to adopt a "What if" mindset. "What if" encourages us to explore possibilities and solutions rather than focusing on obstacles and reasons for failure. This mindset shift is essential for fostering creativity and progress.

Consider the example of the first Google Doc. What if we didn't have a "What if" mindset? We would have seen only the challenges and stopped before even starting. Instead, we asked, "What if this works?" This question drove us to experiment, iterate, and eventually create a product that revolutionized collaborative work.

The Vital Pulse of Innovation

Innovation is the lifeblood of our progress as a species. Every advancement, from the discovery of fire to the development of artificial intelligence, has stemmed from individuals who dared to push against the status quo, who made mistakes, and who refused to accept limitations. These trailblazers saw potential where others saw only obstacles. They transformed dreams into reality through relentless curiosity, unyielding perseverance, and an unshakable belief in the power of human ingenuity.

Throughout history, we see the impact of those who chose to innovate. Imagine a world where early humans, faced with the harsh elements, never experimented with fire. The darkness would have remained unchallenged, and the warmth that fuels our survival might have gone undiscovered. Consider the Wright brothers, whose dreams of flight defied gravity itself. Their innovation transcended the earthbound limitations, giving humanity wings to explore the skies and beyond.

Innovation is not merely a process; it is a testament to the human spirit. It is the refusal to be content with the present and the courage to envision a better future. It is the spark that ignites revolutions, the force that propels us forward, and the light that guides us through uncharted territories.

Embracing the Era of AI with Optimism

In this era of rapid technological change, the call to innovate has never been more urgent or exhilarating. Artificial intelligence presents a frontier rich with challenges, but even richer with opportunities. It offers us tools of unprecedented power, capable of transforming industries, enhancing human capabilities, and solving problems that once seemed insurmountable.

AI is more than just a technological advancement; it is a new lens through which we can view and shape our world. By embracing a mindset of optimism and curiosity, we open ourselves to the vast potential that AI holds. This mindset is not about ignoring the risks or challenges but about approaching them with a solution-oriented perspective. It is about asking "What if" questions that unlock new possibilities and drive progress.

As you move forward, remember that innovation is not a linear path. It is filled with twists, turns, setbacks, and breakthroughs. It is a journey of resilience, where each failure is not an endpoint but a stepping stone to success. Thomas Edison, in his quest to perfect the light bulb, famously said, "I have not failed. I've just found 10,000 ways that won't work." This spirit of perseverance is the essence of innovation.

Stay curious. Curiosity is the engine of discovery. It compels us to ask questions, seek new knowledge, and explore the unknown. It drives us to look beyond the horizon and envision what could be. In a world where AI and technology evolve at a dizzying pace, curiosity is our compass, guiding us to new frontiers.

Stay optimistic. Optimism is the fuel that powers our journey. It gives us the confidence to take risks, the strength to overcome obstacles, and the belief that our efforts will lead to meaningful change. Optimism is not blind faith; it is a strategic mindset that focuses on possibilities rather than limitations.

Most importantly, keep asking, "What if?" This simple yet profound question is the seed of all innovation. It challenges the status quo, disrupts conventional thinking, and opens the door to transformative ideas. "What if?" is the question that led to the moon landing, to the creation of the internet, and to the development of AI. It is the question that will lead us to future breakthroughs that we cannot yet imagine.

Your Turn: Becoming a Creative Entrepreneur

Now it's your turn. Take the ideas from this book and start your journey as a creative entrepreneur in the age of AI. Here are 10 examples from my Sunday Letters, each accompanied by a simple daily practice or exercise to nurture your innovative spirit:

1. **Embrace Failure:** Reflect on a recent failure. Write down what you learned and how it can lead to future success. Celebrate messes! Make virtue from error.
2. **Curiosity Journal:** Spend 15 minutes each day exploring a new topic or technology. Jot down your thoughts and questions. No editor! Just ask "What if? Be silly, stupid, and wrong!
3. **Networking:** Reach out to a new person in your field each week. Share ideas and seek feedback. Don't be afraid to share the ideas that you're unsure of. Brainstorm and look for connections.
4. **Prototyping:** Dedicate time each day to work on a small prototype or experiment. It doesn't have to be perfect; it just has to be started. Small, cheap, and fast is better than long and perfect. See what you can learn in a day, no matter how small.
5. **Mindset Shift:** Start each day by listing three "What if?" questions about your current projects. Try to be as radical as you can be. Look for hidden assumptions and see if you can imagine the world where those assumptions don't exist.

6. **Continuous Learning:** Set aside time each week to learn a new skill relevant to your field. Don't worry about whether it's "valuable" or not. Play with something that interests you. Learn something cool and impractical.
7. **Mentorship:** Find a mentor and/or become one. Share experiences and insights to foster mutual growth. Explaining something is the best way to know if you really understand it.
8. **Collaborative Projects:** Engage in a collaborative project outside your regular work. This expands your perspective and sparks new ideas.
9. **Mindfulness:** Practice mindfulness or meditation to clear your mind and enhance creative thinking. Try to spot your inner editor, the voice of "Why bother" and gently encourage it to sit over there in the corner while you create.
10. **Reading:** Read widely and diversely. Books, articles, and research papers can provide new insights and inspiration. Read for fun. Don't just read things with a purpose. Explore the things that are interesting, even if they don't make sense.

Your Role in Shaping the Future

Thank you for joining me on this journey. Now go forth and innovate. The future is in your hands. Innovation comes from all of us. If you are in the younger generation, you may feel less comfortable making messes and exploring randomly, but this is the time of your life when you have the most ability to challenge the status quo and take those risks. Try to encourage yourself to play and explore whenever you can, and be patient if the ideas don't bear fruit at first. Whether you are an entrepreneur, an engineer, a scientist, or a dreamer, your contributions matter.

Here's my "What if?": Imagine a world where AI enhances our capabilities, amplifies our creativity, and drives us toward a future of abundance and possibility. Where everyone has access to education, healthcare, and creative tools. Where the massive surplus of cognitive energy that AI represents is used to innovate and create in ways that make our descendants' lives seem as unimaginably wonderful as ours would seem to a pre-industrial age aristocrat.

As you embark on your journey, remember that innovation is a collective endeavor. It thrives in an environment of shared knowledge, open dialogue, and mutual support. Seek out mentors, collaborate with peers, and be generous with your insights. Most of all, be gentle with yourself. There is a lot of failure in innovation. Many of those "failures" are successes that have not yet revealed themselves. Don't judge harshly, don't be too impatient. Try, experiment, observe, rinse, and repeat.

The path ahead may be uncertain, but it is also filled with limitless potential. Embrace the challenges, celebrate the successes, and learn from the failures. Let your curiosity guide you, your optimism inspire you, and your "What if" questions drive you. The next great idea could be yours, and it could change the world.

Now, go forth and innovate. The future awaits, and it is yours to shape.

Acknowledgments

I've always thought, I'm glad I don't have to write an Acknowledgments chapter. I'd be sure to miss someone! Well, I guess we'll find out — apologies in advance.

I've learned from many people in my career. Probably everything in this book came from working with someone, and many came from annoying someone. If you're one of the latter, sorry! The first half of my career was spent with my partner in crime, best man at my wedding, founder of 6 companies with me, and the person who gave me my first job, Steve Newman. Steve and Claudia Carpenter were the co-founders of Upstartle (which created Writely), that led to all of the Google Docs stories in this book. I'm sure they remember some of them differently — I tried my best to keep it accurate. Thanks also to Jen Mazzon, our early tester and endless champion.

GDocs itself had many helpers and supporters in and out of Google. Thanks to Nat Torkington at O'Reilly for seeing the early value in it even before we did; Marc Benioff for giving really great advice early on, Eric Schmidt for pounding on the table and insisting Google buy us; Fuzzy, Micah, Jonathan, Steven and the rest of the NYC team who built the incredibly hard spreadsheets features, took over and cleaned up my messes, scaled up, and generally made it what it is today. Thanks to Jeff Huber for being a guiding light and giving a very raw new manager good advice and space to learn. And many, many thanks to the community of Google engineers who helped us get it into that platform, understand all the ins and outs of many different parts of that stack, and educated us in the ongoing cloud revolution. Oh, and Mike Pearson, our M&A person at Google — sorry (he knows why)! It worked out in the end.

There are many folks who have been patient, taken a chance on me, and helped me learn in my career. Thanks to Dan Levin and Aaron Levie for bringing me into Box and furthering my growth as a leader, which led directly to the letters that form the core of this book. Much appreciation to Kevin Scott, a good friend, patient listener, and someone who has always taken time to understand me, sometimes when I didn't understand myself. I am very grateful to Kevin (and Satya) for the chance to be at Microsoft, learning about this new land of AI and helping expand our reach in the process. Many Microsoft folks have also made me feel at home and helped me learn this new (to me) culture: Eric Horvitz, Rajesh Jha, Peter Lee, Jaime Teevan, Dee Templeton, and so many more (see! I know I'm forgetting folks). Thanks to Tim Bozarth, one of the best system thinkers I know, who has taught me how to look through the lens of incentives more clearly.

Thanks to Greg Shaw, editor at 8080 Books, for believing in this book and pushing it into existence. Thanks also to the 8080 team, including Steve Clayton and Katie Oakes Stevens as well as the folks at Partners Media, Myles Thompson, Leah Lusk, and Barbara Byrne.

When I was 12, my father bought me an early computer (TRS-80). I barely understood why I wanted it, I can't imagine he knew why he was spending so much money on it. But he did, and I've always been grateful for that. It's a great example of "what if," and it paid off, and started me down this road. And thanks, mom, for seeing and believing in my math skills when early high school teachers couldn't see it. That got me where I needed to be in a time when I needed it most.

Thanks to James Cham, who has listened to, amplified and encouraged many strange ideas over many coffees.

Finally, thanks to my ever-patient wife Angela, and my kids Dominic and Sofia. "Dad makes messes and has strange ideas! He holes-up in the office a lot. He works on things that don't make sense. We've never been able to park a car in the garage because of his shop!" Thanks for all the love, patience, space, and understanding to learn what I've learned here. It is indeed a rollercoaster, and I am so happy and proud that each of you are finding your own creative paths in the world.

About the Author

Sam graduated from the University of Michigan with two math degrees and has spent the past 35 years in Silicon Valley. He is as a six-time founder of consumer and enterprise startups, including the startup that created Google Docs. He worked at Google off and on since 2006, at some point running many of the company's consumer apps: Gmail, Blogger, Picasa, GDocs, Page Creator, Reader, and parts of Maps. He has been a venture investor at Google Ventures, and was the SVP of engineering at Box leading up to its IPO. He now works in the office of the CTO at Microsoft, where he works on a number of projects related to product development culture and practices, including a large horizontal infrastructure effort, consumer design culture initiatives, and a number of AI projects, including the Semantic Kernel app building framework.

www.ingramcontent.com/pod-product-compliance
Lightning Source LLC
LaVergne TN
LVHW020726171224
799306LV00025B/904